WAR 4.0

ARMED CONFLICT IN AN AGE OF SPEED,
UNCERTAINTY AND TRANSFORMATION

WAR 4.0

ARMED CONFLICT IN AN AGE OF SPEED,
UNCERTAINTY AND TRANSFORMATION

**EDITED BY DEANE-PETER BAKER
AND MARK HILBORNE**

Australian
National
University

ANU PRESS

Australian
National
University

ANU PRESS

Published by ANU Press
The Australian National University
Canberra ACT 2600, Australia
Email: anupress@anu.edu.au

Available to download for free at press.anu.edu.au

ISBN (print): 9781760466817
ISBN (online): 9781760466824

WorldCat (print): 1499172797
WorldCat (online): 1499175010

DOI: 10.22459/W4.2025

Cover design and layout by ANU Press

This book is published under the aegis of the Asia-Pacific Security Studies editorial board of ANU Press.

Contents

Introduction

Deane-Peter Baker and David Pfotenhauer

> In war, science has proven itself an evil genius; it has made war more terrible than ever before. Man used to be content to slaughter his fellowmen on a single plane—the Earth's surface. Science has taught him to go down into the water and shoot up from below and go up into the clouds and shoot down from above, thus making the battlefield three times as bloody as it was before; but science does not teach brotherly love. Science has made war so hellish that civilization was about to commit suicide; and now we are told that newly discovered instruments of destruction will make the cruelties of the late war seem trivial in comparison with the cruelties of wars that may come in the future.[1]

These words were part of the summation presented by William Jennings Bryan in July 1925 at the conclusion of the Scopes Trial (*The State of Tennessee v. John Thomas Scopes*), which has gone down in history as a major turning point in the social acceptance of the theory of evolution. On the day, Bryan was on the winning side—John T. Scopes, the high school teacher who had, in defiance of the law, dared to teach human evolution in a state-funded school, was convicted and fined (the conviction was later overturned on a legal technicality). The victory was, however, pyrrhic. The attention it drew bolstered the pro-evolution cause, which eventually won the day.

Contrary to the caricatured depictions of Bryan that have since emerged (most notably as the character Matthew Harrison Brady in the 1955 play *Inherit the Wind* by Jerome Lawrence and Robert Edwin Lee, and as the Cowardly Lion in L. Frank Baum's *The Wonderful Wizard of Oz*), he was a complex character. A three-time Democratic Party presidential nominee

---

1 William Jennings Bryan quoted in Marvin Olasky and John Perry, *Monkey Business: The True Story of the Scopes Trial* (B&H Books, 2005), 325.

and secretary of state under Woodrow Wilson, Bryan combined his deep Christian faith with a (largely) progressive social agenda. As Scott Farris writes in *Almost President: The Men Who Lost the Race but Changed the Nation*, 'he occupies a rare space in society … too liberal for today's religious, too religious for today's liberals'.[2]

While the vast majority of people today would disagree with Bryan's views on the evils of evolutionary theory, his depiction of science as 'the evil genius of war' is an image that would likely resonate with many. Perhaps nowhere is this more evident than in the debate over lethal autonomous weapons (LAWS). The civil society effort to regulate such systems is, at the time of writing, dominated by a coalition known as The Campaign to Stop Killer Robots. The ingenious title for 'The Campaign' (as it is generally known) taps into widespread and deep-seated fears. Fed by images from Hollywood movies such as the *Terminator* series, the idea of a 'killer robot' is one we are intuitively predisposed to oppose. And it is not just fear that is a driving force—at the core of the resistance to the development and use of LAWS is, for many, a powerful intuition that this is a moral threshold that should not be crossed. Here the opening lines of Bryan's summation resonate deeply:

> Science is a magnificent force, but it is not a teacher of morals. It can perfect machinery, but it adds no moral restraints to protect society from the misuse of the machine.[3]

Though the phrasing of his words is a little dated to our ears, Bryan's summary of the impact of science on war is prescient. Where we once waged wars on only one 'plane', science has now enabled us to do so in multiple domains. In 1925, the year of the Scopes Trial, William Lendrum 'Billy' Mitchell published his ground-breaking treatise *Winged Defense: The Development and Possibilities of Modern Air Power—Economic and Military*. Mitchell is now widely recognised as one of the leading visionaries in developing an understanding of the implications of airpower. Yet even Mitchell did not predict that the warfighter's reach would one day extend into space, as it now does. And neither Bryan nor Mitchell could have foreseen that we might create an entire artificial 'plane'—cyberspace—which would become another domain for war. Nor could they have imagined in 1925 that science would, through the development of nuclear weapons, lead civilisation to the very brink of suicide. As is now well known, it was only the decision by

---

2   Scott Farris, *Almost President: The Men Who Lost the Race but Changed the Nation* (Lyons Press, 2011), 93–94.

3   Bryan quoted in Olasky and Perry, *Monkey Business*, 324.

Lieutenant Colonel Stanislav Yevgrafovich Petrov to ignore (against orders and protocol) the alarm raised by the Soviet Union's nuclear attack early warning system that prevented nuclear Armageddon in 1983.

Science, of course, has also reaped untold benefits for humanity, from the field of medicine to communications and engineering and beyond. And even in the realm of war, science has not trended unambiguously in the hellish and cruel direction that Bryan predicted. Consider, for example, the scientific advances that have enabled the development of precision-guided munitions. As the Royal Australian Air Force Air Power Development Centre has noted:

> In World War II, it would take 108 B-17s dropping 648 bombs to get two bombs onto an intended target which by the 1991 Gulf War could be achieved by a single aircraft using precision guided munitions, if the prevailing conditions were right.[4]

Of course, the technical ability to achieve *precision* is not the same thing as the ethical imperative to engage targets with *discrimination*, but precision does enable discrimination. This highlights—as does the example of Stanislav Petrov above—the reality that even technology-enabled war remains a human endeavour, and one for which humans remain responsible. The connection between the human and technological elements of war, in the rapidly changing conflict environment that is emerging and developing, is the central uniting theme of the chapters in this book.

It is widely agreed that we find ourselves enmeshed in a fourth industrial revolution. Each of these revolutions has carried with it significant evolutions in military technology. The advent of steam power enabled the development of steel ships and submarines. Electricity underpinned the mass production of the second industrial revolution which led to an exponential growth of military capabilities. The third industrial revolution, driven by the development of electronics and information technology, radically altered military command, control and targeting. Now, as World Economic Forum founder Klaus Schwab puts it:

---

4   RAAF Air Power Development Centre, 'The Accuracy of Air-Delivered Weapons', *Pathfinder: Air Power Development Centre Bulletin* 342 (October 2019): 2.

a Fourth Industrial Revolution is building on the Third, the digital revolution that has been occurring since the middle of the last century. It is characterized by a fusion of technologies that is blurring the lines between the physical, digital, and biological spheres.[5]

Schwab offers three key reasons why we should not think of our current era as merely an extension of the third industrial revolution:

> velocity, scope, and systems impact. The speed of current breakthroughs has no historical precedent. When compared with previous industrial revolutions, the Fourth is evolving at an exponential rather than a linear pace. Moreover, it is disrupting almost every industry in every country. And the breadth and depth of these changes herald the transformation of entire systems of production, management, and governance.[6]

While these changes are to some degree still more promise than practice, their potential is massive. We are only now beginning to come to grips with the implications this will have for the conduct of warfare. Though necessarily only a starting point, this book offers the perspectives of a collection of leading international experts addressing aspects of this new era of war—what we might think of, broadly, as War 4.0.

It's important to note that the chapters of this volume were mostly written prior to the Russian invasion of Ukraine in 2022 and the war in Gaza that commenced in 2023. While those conflicts have thrown up a range of interesting issues, on reflection the editors of this volume judged that seeking to account for these conflicts through revisions to the chapters would have added insufficient benefit to warrant delaying publication.

War, technology and society are, and always have been, inextricably linked. In this volume we address a range of individual technologies including artificial intelligence, unmanned aerial vehicles, autonomous weapons and quantum computing. But the true impacts of these capabilities can only be understood in the context of the overarching societal and demographic shifts that are emerging. In the first chapter of this volume, Brigadier Ian Langford examines the changing system of global order and how these changes intersect with the rapid, disruptive effects of new technology. The chapter maps the various actors, physical features, trends and technological

---

5   Klaus Schwab, *The Fourth Industrial Revolution: What It Means, How to Respond* (World Economic Forum, 14 January 2016).
6   Schwab, *The Fourth Industrial Revolution.*

variables that will shape the future operational environment. Issues such as urbanisation, population shifts, climate change and changing state behaviours are examined within the broader scope of technological democratisation. In particular, Langford examines the emerging trend towards the obsolescence of state domination, as a result of technological disruption. He argues that the previous global system, characterised by state-centrism, is moving towards a multilevel, multinodal system. Within this evolving global architecture of order, the democratisation of technology has democratised warfare. Langford argues that a troubling outcome of the democratisation of technology and its impact on war is that the technological superiority of state-based militaries over sub-state actors is being offset by technological democratisation. Langford also addresses the issue of the sheer complexity of a future operating environment populated by a bewildering array of interconnected variables including mass urbanisation, resource scarcity, population shifts and climate change. Coupled with this, Langford ties in the current and future impacts that artificial intelligence, robotics, space and cyber security will have on the manner in which states employ these capabilities in an operating environment that is increasingly crowded and complex. The chapter ends on a cautionary note by affirming the expectation that states will still be the dominant actors of the foreseeable future—but that domination will be sorely tested by an operational environment that is increasingly democratised, dispersed and complex.

In Chapter 2, Nadya Bliss addresses the challenge to democratic Western governments in reclaiming their role as the primary global engines of technological innovation. Bliss reviews the initial, key role that the United States played in technological innovation and the subsequent decline of this role in the face of increasing private sector domination. The overarching theme of this chapter is that the reduction of governments' role in technology research and development has impacted negatively on national security. Bliss advocates for a rejuvenation of the US innovation ecosystem in which a mutually beneficial relationship existed between the federal government, higher institutions of learning and the private sector. This, argues Bliss, will result in a more resilient response to the use of commercial technology as a disruptive force by adversaries. The partnership between federal research and development and the private sector has, according to Bliss, proven to be an impactful and effective model of joint technological innovation. She highlights that one area of opportunity for governments to reclaim their important role as drivers of innovation is in combating disinformation for political purposes. As an example, the chapter employs the manipulation of social media companies like Facebook and Twitter as weapons used by Russia to disrupt and manipulate domestic politics during the 2016 election year in the US. Bliss argues that because the innovation ecosystem in the US was not effectively balanced between government and private sector actors, both were unprepared to combat such unique forms of interference and disruption.

David Kilcullen follows on from Bliss's account by addressing the continuing challenge experienced by governments in preparing for future war. Kilcullen points us to an operating environment that is cluttered, highly connected, crowded, predominantly urban and coastal, and filled with a mix of state and non-state actors. The sheer immensity of the hundreds of variables that span the political, military, economic, social, infrastructure, informational and physical contexts generate severe problems of overload. In this environment, both state and non-state adversaries are employing unorthodox methods aimed at overwhelming state-based forces with a massive number of small challenges rather than a single catastrophic threat. Kilcullen leverages the description of the post–Cold War era by former CIA director Jim Woolsy as one in which the dragon has been slain but is now beset by snakes. Consequently, argues Kilcullen, the usual means by which we may attempt to project the character and form of future war provide little clarity. He argues that governments' quest for reliable predictions must, in the face of a complex and unpredictable operating environment, be replaced with the recognition that any assessment can offer only a hypothesis of future trends. Kilcullen notes that irrespective of the complexity of the numerous variables impacting the character of future war, they can be usefully categorised into a simpler taxonomy of constants (features), trends (continuities) and shocks (discontinuities). The most important predictive variable in the taxonomy for analysing future conflict, argues Kilcullen, is shocks. Focusing on potential discontinuities, as opposed to emphasising constants and trends, will, he argues, create improved resilience for the inevitable impact of shocks on policy, strategic planning and doctrine. Kilcullen illustrates his contention by way of the example of the erroneous Cold War–era belief in the persistence of the Soviet Union, as well as the misguided expectation that China would conform to past patterns of economic growth and military–political development. Urging caution but arguing for proactive measures to plan, as best as possible, for the challenging operating environment of tomorrow, Kilcullen again reminds us of the need to be wary of orthodoxy and instead embrace innovative planning models.

In the fourth chapter, Dr Mark Hilborne focuses on the domain of space, which is only now beginning to emerge in the security estimations of many states, though it has been central to both military capabilities and increasingly to the functioning of modern economies. An inherent difficulty of the domain is the dual-use nature of much of the relevant technology. Added to the general ambiguity of space, this makes distinguishing military operations and capabilities from non-military difficult, adding complexity to efforts to secure space. The growing commercial sector is itself changing the complexion of the domain and beginning to challenge military forces as the largest supplier of data and communications. China's opaque policy-making procedures obscure the demarcation between their commercial, civilian and military sectors—made manifest by the policy of Military–Civil Fusion (MCF). These dynamics create uncertainty, and wariness from the US, whose investment in, and dependence on, space has become critical, increasing the likelihood of misperception and tension.

In the fifth chapter Malcolm Davis focuses in on a key emerging warfighting domain. Space, Davis contends, is no longer a peaceful, neutral domain—it is a highly militarised environment, filled with an assortment of technology that directly supports war-based activities on Earth, and which is likely to be weaponised in the near future. Davis cites China's test of an anti-satellite weapon in 2007 as a watershed event in reframing the debate over space as a domain of war. Davis urges that understanding this new frontier of warfare is essential if governments are to effectively plan for the impact of space as an emerging strategic domain. Davis posits that the next major war is likely to start in space—and in cyberspace, another new warfighting domain, one which is increasingly linked with space operations. Throughout the chapter, Davis reflects on how this marriage between emerging space warfare thinking and current cyber capabilities relates to current US and Australian planning for the risks associated with space warfare more generally. Davis links terrestrial conditions of conflict—such as 'contested, congested and competitive'—with space, and argues that to cope with this similarity, the emphasis must be on defensive counterspace resilience rather than offensive counterspace capability.

We cannot, of course, meaningfully plan to compete and potentially fight in space without understanding the parallel efforts of likely adversaries. In Chapter 6, Jian Zhang examines the Chinese People's Liberation Army's (PLA) view of space warfare and its efforts to develop Chinese space capability, strategy and operational doctrine. In the first part of the chapter, Zhang examines the factors driving the increased emphasis on space in China's national military strategy. He notes that there has yet to be a formalisation of space strategy or space operation doctrine; however, he reveals and analyses a rich vein of current discussion on emerging principles of space strategy and doctrine which can be found in key government publications. Noting that, after the US, China has the world's largest satellite system, Zhang goes on to discuss the PLA's current efforts to produce a range of kinetic and non-kinetic counterspace capabilities, including missiles, electromagnetic and cyber weapons, directed-energy weapons and co-orbital satellite killers. Zhang's account reveals that China's space warfare thinking and emerging military capabilities demonstrate a cautious yet firm approach to the operational and strategic requirements of space warfare.

Turning from outer space to inner space, in Chapter 7 Marcus Doherty provides us with a vision of how quantum technologies might impact the future of war. He qualifies his account by pointing out that while these technologies have certainly moved beyond being mere 'scientific speculation', we are still some way from understanding the full range of applications and impacts of this new technological frontier. After providing a definition, and overview of the shared characteristics of quantum technologies, Doherty provides a survey of the currently understood applications of quantum technology. These include quantum-based sensing and imaging capabilities, quantum communications and cryptology, as well as quantum computing and simulation. Doherty then explores the most immediate likely applications of these technologies in the defence context, before providing a vision of the impact—opportunities and risks—of the most likely cumulative evolution of quantum technologies over the next 20 years.

Though technology will unquestionably be the key driver of change to military operations in the emerging future, humans will remain intrinsic to the warfighting endeavour. In Chapter 8, drawing lessons from the UK Royal Air Force experience of employing the Reaper drone, Peter Lee reflects on the human impact associated with current technologies that distance humans from the physical acts of conflict and war. Lee argues that despite the distance created by emerging technologies in remote warfare, there is still a profound degree of intimacy associated with the act of killing remotely—what he refers to as 'the distance paradox'. While the human operators are geographically far away from their victims, the crews witness and experience events of killing in great detail. Lee builds this narrative around a brief historical overview which demonstrates how physical and psychological distance in weapon use by air, land and maritime forces have increased in tandem over time. The subsequent sections explore the potential effects of this visual intimacy and the re-humanising of targets on operators, introducing physical, emotional and psychological responses. Moral injury is also discussed in relation to attempting to understand why some Reaper crews last longer, in terms of operational tours, than others. The final section of the chapter considers questions about the costs and benefits of empathy with human targets and the extent to which the previous factors might contribute to its occurrence or prevention.

One potential solution to the risk of moral injury and related harm to crews of remote weapons systems is to take human out of the loop and instead make these systems autonomous. The prospect (and reality) of LAWS is, however, highly controversial. In Chapter 9, Ian MacLeod and Erin Hahn provide a perspective on how we might further the international debate on the moral and ethical issues associated with the use of LAWS in war. MacLeod and Hahn observe that at the heart of the discussion regarding the use of LAWS in war is the question of whether the development and fielding of LAWS inappropriately removes the human role in the decision to kill, and if so, whether the consequences (operational, moral, legal) are acceptable. They argue that the debate over LAWS is deficient, primarily because it fails to take into account the way military forces make operational targeting decisions in practice. Additionally, the language and terminology routinely used in arguments about LAWS, and artificial intelligence more broadly, anthropomorphise the technology in a way that conveys a flawed and superficial understanding of these emerging capabilities. The framing deficiency is addressed by MacLeod and Hahn by introducing an operational lens through which to evaluate existing weapons systems and example cases of LAWS.

Because war without ethics is just murder writ large, the final chapter is reserved for an evaluation of whether the 'just war' tradition is sufficient to address the complexities of future war. Valerie Morkevičius contends that, while on the surface the classical tradition may seem to be outmoded and not fit for purpose, no matter what future warfare looks like, going back to the past is a useful tool to look into the future. In doing so, Morkevičius brings to our attention seven key lessons from the past that she argues will remain relevant to future war: there never has been a golden age of war, war is endemic to the human experience, war will inevitably cause harm to the wrong people, death is not the worst outcome of war, ethics in war is not reducible to simple calculations and ethical action in war will always demand good character. The final, and perhaps the most important, lesson

Morkevičius draws from the past is that we must recognise that our own parochialism—our views about war, and the ethics of war—are inescapably embedded in a time and a place. In this she is united with all of the contributors to this book who, each in their own way, challenge us to adopt a stance of humility and openness as we seek to understand and respond to War 4.0.

References

Farris, Scott. *Almost President: The Men Who Lost the Race but Changed the Nation.* Lyons Press, 2011.

Olasky, Marvin, and John Perry. *Monkey Business: The True Story of the Scopes Trial.* B&H Books, 2005.

RAAF Air Power Development Centre. 'The Accuracy of Air-Delivered Weapons', *Pathfinder: Air Power Development Centre Bulletin* 342 (October 2019).

Schwab, Klaus. *The Fourth Industrial Revolution: What It Means, How to Respond.* World Economic Forum, 14 January 2016. www.weforum.org/agenda/2016/01/the-fourth-industrial-revolution-what-it-means-and-how-to-respond/.

1

Accelerated Change—The Evolving Character of Society and Conflict in an Age of Speed, Uncertainty and Transformation

Ian Langford

Introduction[1]

The 21st century is being shaped by a range of exciting, worrying and accelerating trends and technologies: robotics, artificial intelligence, 5G networks, cyber threats, social media, 3D printing, bio-enhancements and a new geopolitical competition. These trends and technologies will also affect the character of world conflict and warfare in the future.

Recent years have seen a change in the international security environment. There has been a shift of economic and military power from the Atlantic towards the Pacific. Part of these economic power shifts is the relocation

---

1   Editors' note: as mentioned in the Introduction, this chapter was written in 2020, and global events since then may have overtaken some aspects. As the pace of technological and cultural change has continued to accelerate, the editors have opted to present this text as drafted to minimise further delays in publication.

of the world economy's centre of gravity towards Asia, where it has been estimated, for example, that the size of China's economy will surpass that of the US by 2028.[2]

By 2035, the Asia-Pacific region will resemble a dual hierarchy where security and economic superiority will be fundamental to the competitive strategies of both the US and China.[3] Competitive behaviour appears to be inevitable as relative geostrategic power is challenged. To put it in perspective, the cost of the Belt and Road Initiative is almost twice that of the Marshall Plan following the Second World War but is being realised through a system of leveraged loans rather than through the gifting of grants.[4] This gradual increase in political influence is in contrast with the failing notion of the 'West' as a group of countries, led by the US, defined by core values since the end of the Second World War.[5] The shift in geostrategic power balance coupled with a distracted European Union and increasingly inward-looking US, has provided opportunities for nations such as Russia and North Korea to further their interests. This has manifested in punitive gains and grandstanding that are resulting in a loss of confidence in historic systems of stability and trusted institutions, including NATO (the North Atlantic Treaty Organisation), the United Nations and the World Bank.

Most Western commentators and academics agree with Carl von Clausewitz's maxim that the fundamental nature of war does not change, merely its character.[6] Military forces the world over have had to adapt to a range of diverse challenges since the terrorist attacks of September 11, 2001: peacekeeping, peace enforcement, counterinsurgency, medium-intensity warfighting, counterterrorism and humanitarian assistance/conflict termination. The concept of purely conventional warfare in the 21st century as being a first resort of nation-states to coercively achieve national interests is no longer relevant. Most adversaries now engage in fringe unconventional activities (often termed 'grey zone' warfare) to weaken the influence and

---

2   Dan Steinbock, 'The Global Economic Balance of Power Is Shifting', *World Economic Forum*, 20 September 2017, www.weforum.org/agenda/2017/09/the-global-economic-balance-of-power-is-shifting. These calculations were prior to the current economic slowdown, which will likely affect this prediction, but it remains indicative of the importance of China.
3   NATO, *Strategic Foresight Analysis Report 2017* (Norfolk: NATO Strategic Analysis Branch, 2017), 23.
4   Anja Manuel, 'China Is Quietly Reshaping the World', *The Atlantic*, 17 October 2017, www.theatlantic.com/international/archive/2017/10/china-belt-and-road/542667/.
5   Daniela Schwarzer, 'Europe, the End of the West and Global Power Shifts', *Global Policy* 8, no. S4 (2017): 18–26, onlinelibrary.wiley.com/doi/full/10.1111/1758-5899.12437.
6   Carl von Clausewitz, *On War*, ed. and trans. Michael Howard and Peter Paret (New York: Knopf, 1976).

power of international rivals and undermine civil societies because they know that asymmetric options capable of exhausting and defeating nations as preferable to open warfare.

Aggressive nations and hostile subnational groups, such as the Islamic State of Iraq and Syria (ISIS) now deliberately avoid conventional warfare. This trend is likely to continue. Rival states, such as Russia and China, as counterweights to US strategic military power, are adopting hybrid warfare strategies as well as employing high technology capabilities such as cyber, robotics and drones. It is almost as if Edward Luttwak's paradoxes of strategy (2001) are showing empirically now as the West's greatest strength— its collective conventional military dominance on the ground and sea, and in the air—is evaded by hostile national and subnational groups, resulting in Western military forces unable to effectively operate against insurgents, terrorist networks and state-sponsored militias in situations now classified as essentially hybrid and grey zone in their character.[7]

This chapter considers the impact of a changing global system which is increasingly being defined by notions of state-based competition and conflict, as well as the rapid, disruptive effects of new technologies. These technologies are shaping the way that people live, how they communicate and learn about the world, how they consume resources and access services, and how they enable their militaries to fight. Factors informing the competition that we currently see include urbanisation and population shifts, technology proliferation and adaptation, climate change and changing state behaviours within the international system.

Regional developments

The Asia-Pacific region has benefited from a strong US presence and security guarantee since the end of the Second World War; increasingly, however, major global powers are now vying among themselves for greater influence. As the security environment increases in complexity, so too does the risk of conflict. With growing economic power, many regional militaries are seeking to rapidly modernise to meet this perceived risk, as seen by the steady growth in defence spending for the last 30 years. This trend is expected to

---

7   Jahara Matisek and Ian Bertram, 'The Death of American Conventional Warfare: It's the Political Willpower, Stupid', *The Strategy Bridge*, 5 November 2017, thestrategybridge.org/the-bridge/2017/11/5/the-death-of-american-conventional-warfare-its-the-political-willpower-stupid.

continue while regional countries' economic circumstances strengthen and the geopolitical environment remains uncertain. While defence expenditure is traditionally low in the Pacific Island countries, with some only having the capacity for police forces for law and order duties, increasing assistance is being provided by external military partners, including Russia, China, the US and Australia.

China's strategic interests appear to be to gain increased support among international bodies, as well as to deepen its ties and influence throughout South-East Asia and the south-west Pacific. This is achieved by developing fishing ports and maritime zones, as well as securing future access to seabed mining and increased political influence among Pacific Island countries. China has demonstrated its intentions through recent observable growth in trade, investment and physical presence, a trend which is likely to continue.[8]

Military dominance of the US in the Indian-Pacific and NATO in the West should no longer be assumed.[9] The effect of the shift in power between the US and emerging global powers such as China and India is likely to result in an increased sense of 'competition' as a new norm, especially as some nation-states increasingly respond to the change in the international systems by focusing their foreign policy and security settings on 'interests' rather than 'values'.

There is little to suggest that Australia will face a physical existential threat in the near term. Noting the complexity of regional interactions, however, there is a risk that Australia's traditional advantages are at risk of being undermined through a loss of influence. This presents an ongoing challenge, noting that Australia's population growth and commensurate skilled workforce will be insufficient as a means to continue to secure its interests in an increasingly networked global system.

Urbanisation and overpopulation

The world is experiencing an asymmetric growth in population, which has a number of projected follow-on effects, including competing demands for finite resources. The developed nations are aging and their growth is dependent upon net migration, whereas less developed countries are experiencing high

---

8   New Zealand Foreign Affairs and Trade, *Pacific Diagnostic 2015–2035* (Wellington: NZ MFAT, 2015), 46.
9   NATO, *Strategic Foresight Analysis*, 21.

rates of growth yet are significantly younger. Asia[10] is, and will continue to be, dominated by China and India. India is expected to overtake China by 2025 with a population of 1.48 billion. While there are questions over the reliability of official Chinese statistics, indications are that China's population growth is low or in decline;[11] however, India, and much of Africa and South-East Asia, are experiencing growth rates of over 10 per cent. These rates of growth are expected to exceed most of the host nation's projected infrastructure and resource capacity. Noting that working-age populations are strong indicators of national power (though there are certainly other important considerations), India is likely to increasingly compete for its share of regional influence with countries such as China and the US.

The Asia-Pacific region is highly likely to see continued growth of 9–13 per cent. The projected population of the Oceania region is 46.65 million while South-East Asia is expected to grow to 717.4 million by 2028. The distribution of the population is likely to, in the absence of a catastrophic event, remain similar to today. Australia, Papua New Guinea (PNG) and New Zealand will continue to dominate Oceania with 92 per cent of the total population. Likewise, Indonesia should remain the largest country with 291.5 million and, based on the nearest competitors of the Philippines (122.3 million) and Vietnam (105 million), is expected to remain so for the foreseeable future.

Table 1.1: Future urbanisation of global population.

Region	Urban (millions)		Rural (millions)	
	2015	2050	2015	2050
Africa	492	1,489	703	1,039
Asia	2,120	3,479	2,300	1,778
Europe	547	599	194	117
Latin America	505	685	127	95
Northern America	291	387	65	48
Oceania	27	41	13	16
WORLD	3,981	6,680	3,402	3,092

Source: UN 2018 Revision: World Urbanization Prospects.

---

10 For the purpose of this study, Central Asia (e.g. Kazakhstan and Uzbekistan) and West Asia (e.g. Iraq, Saudi Arabia and Armenia) have not been included.

11 See, for example, Xiujian Peng, 'China's Population Is Now Inexorably Shrinking, Bringing Forward the Day the Planet's Population Turns down', *The Conversation*, 19 January 2023, theconversation.com/chinas-population-is-now-inexorably-shrinking-bringing-forward-the-day-the-planets-population-turns-down-198061.

Megacity/city-states

It is estimated that urbanisation will increase from 55 per cent to 68 per cent by 2050.[12] Significantly, there will be a number of emerging 'megacities' (cities with populations in excess of 10 million) in the region.[13] As megacities concentrate resource usage and economic output, their relative significance is likely to rise rapidly, particularly as the percentage of populations in slums commensurately increases to about 40 per cent of the urban population. Prevailing conditions in slums create serious economic, social, political and physical insecurities for the inhabitants. Inequality has become a major emerging security challenge.[14] Megacities, and cities in general, now pose significant challenges for most modern militaries.

Imbalances across regions and countries are likely to exacerbate existing political and social tensions and related governance challenges.[15] For example, declining birth rates and longer life expectancy in South-East Asia will create a shift in the demographic profile to an older population. Traditionally the demands of the elderly have been able to be absorbed within social and familiar structures. The increased aging population may lead to intergenerational tensions as the social and political structures necessary to enable their welfare are yet to be implemented and resourced across the region.

There is however a short period of time where the ratio of people of an economically productive age (15–65) supporting the remainder of the population will increase in several countries, presenting an opportunity for economic benefit. For instance, most of Oceania (less Australia and New Zealand) will likely experience a youth bulge.[16] This dividend will be dependent on sufficient opportunity for productive engagement of the labour force to ensure opportunity is realised. With over 50 per cent of the

---

12   United Nations, *World Urbanization Prospects: The 2018 Revision* (New York: United Nations, 2019), population.un.org/wup/.
13   Tal David, 'Our Future Is Urban: Future of Cities P1', *Quantum Run*, 15 September 2020, www.quantumrun.com/Prediction/our-future-urban-future-cities-p1.
14   NATO, *Strategic Foresight Analysis*, 38.
15   United Kingdom Ministry of Defence, *Global Strategic Trends: Out to 2045* (London: Ministry of Defence, 2018), 4.
16   United Nations Population Fund, *Population and Development Profiles: Pacific Island Countries* (Suva: UNFPA, 2014), pacific.unfpa.org/sites/default/files/pub-pdf/web__140414_UNFPAPopulation andDevelopmentProfiles-PacificSub-RegionExtendedv1LRv2_0.pdf.

Pacific Islands countries' population under the age of 25, the region has a surging working-age population but currently lacks sufficient education and employment opportunities to harness this demographic dividend.[17]

A growing number of migrant peoples may look to move between and within Asia and Africa, with Asia becoming an increasingly important destination. Temporary large displacements due to crises will probably continue to occur with high local impact—and the magnitude of such events is likely to be amplified by increased population density. Millions of people may be 'trapped' in vulnerable areas because of the high costs of migration, unable to raise the capital needed for moving away.[18] Unless there is sufficient opportunity for bettering their lives, the urban poor are likely to become frustrated—and with increasing access to information, be aware of growing inequality. If not dealt with effectively, this could lead to violent protests and possibly full-blown urban insurgencies. Access to these urban areas is likely to be highly competitive, with other state and non-state organisations aiming to have freedom of action to base, manoeuvre and operate there.

Technology proliferation and adaptation

The pace of technological change continues to remain high and is shaping how societies, and their military forces, operate. These changes are apparent in electromagnetic sensing, artificial intelligence (AI), robotics and autonomous systems, information warfare, space and cyber developments. As a tactical example, the US military force that went into Afghanistan after the 2001 attacks used zero robotic systems; now the force has over 22,000 in its inventory.[19]

Algorithms, and the workforce of qualified people who design them, will likely be the limiting factor rather than processing power. The ability to process information gets exponentially better and cheaper, and should continue to do so. AI, virtual and augmented reality (VR/AR), drones and unmanned systems, additive (3D) manufacturing, bio-engineering and the internet of things (IoT) have all recently come to public attention despite all having had a long evolution. What appears to be occurring is both a

---

17   New Zealand Foreign Affairs and Trade, *Pacific Diagnostic 2015–2035*, 27.
18   UK Ministry of Defence, *Global Strategic Trends: Out to 2045*, 7–8.
19   PW Singer, *Wired for War* (Boston: Houghton Mifflin Harcourt, 2009); April Glaser and Rani Molla, 'The Number of Robots Sold in the U.S. Will Jump Nearly 300 Percent in Nine Years', *ReCode*, 3 April 2017.

convergence of complementary technologies and a public sector willing to invest large amounts to commercialise these technologies. This is resulting in a changing operating environment. The 'dual use' utility of these technologies will have an obvious technology benefit for military forces.

Developing technologies — Global examples

Distribution of electromagnetic sensing. The ability to hide any power transmission is getting more and more difficult. The confluence of sensor and power source miniaturisation, meta and nano-materials, and an increase in communication networks creates a substantial decrease in detection thresholds. As the reliance on the electromagnetic spectrum increases, the ability to detect emissions is also increasing. The ability to find things is rapidly increasing due to space-based surveillance, networked advanced radars, multipurpose drones and a vast array of sensors that are far cheaper than the technology and techniques that defeat them.[20] Consequently, the capacity to hide now requires the ability to monitor and mitigate signatures beyond the physical to the electromagnetic.

The rapid proliferation, strength, fidelity and range of sensors have created the potential for cognitive overloading of decision-makers. Commanders, and their staff, are facing a situation where they are likely to have sufficient information to make a decision, however, are incapable of analysing the sheer volume of information fast enough. In these situations, commanders are more likely to revert to simple intuition-based decision-making that may not suit the complexity of the operating environment. The increasing sophistication of decision support systems should alleviate some of these problems, provided the commanders' decision-making cycle can be conducted faster than the adversaries'.[21]

Artificial intelligence. The field of AI spans all the way across the computerisation of human decision-making, from machine learning to established neural networks. Arguably no other technology area is seeing

---

20   US Army TRADOC G2 Mad Scientist Initiative, 'An Advanced Engagement Battlespace: Tactical, Operational and Strategic Implications for the Future Operating Environment', *Small Wars Journal*, 24 October 2017, archive.smallwarsjournal.com/jrnl/art/advanced-engagement-battlespace-tactical-operational-and-strategic-implications-future.
21   Jasmin Diab and Nathan K Finney, 'Whither the Hover Tank? Why We Can't Predict the Future of Warfare', *Modern War Institute*, 6 December 2018, mwi.westpoint.edu/whither-hover-tank-cant-predict-future-warfare/.

as much energy and investment in the world today. Currently, there is approximately $153 billion in spending globally, with an estimated annual creative disruption impact of $14–33 trillion dollars.[22] AI is set to massively compound this problem with the creation of 'deepfakes'. Artificial networks can now mimic how the human brain works by having individual nodes that activate to a single point of information and carry out incredibly complex tasks by layering the connections together. Through this, machines now can study a database of images, words and sounds to learn to mimic a human speaker's face and voice almost perfectly. An early example of the potential political impact of this came in the creation of an entirely fake conversation between Barack Obama, Hillary Clinton and Donald Trump.[23] The potential use of this technology for information operations, including misinformation to shape popular opinion and political outcomes, is obvious and deeply concerning.

AI now enables governments, their militaries and businesses to make more accurate decisions via predictions, classifications and clustering. AI accomplishes these tasks by finding patterns within complex unstructured data such as images, audio and documents, or within structured historical data. AI can not only accomplish repetitive, tedious and low-skilled tasks but also read millions of documents, images and data in seconds and extract all the useful information from them faster and with fewer errors than human beings. The 'dual use' potential of AI for both the military and civil sectors of the international system seems almost limitless. As was observed by a lead academic at the Massachusetts Institute of Technology: 'Every area of life will be affected by AI. Every. Single. One.'[24]

Robotics and autonomous systems. The previous decade has shown an exponential trend in foundational technologies such as computing, data storage and communication systems.[25] A convergence of these technologies has resulted in the development of cloud services, machine learning and robotics. While levels of autonomy are already employed in the operating environment, the next decade can expect to produce a maturation of

---

22 'Artificial Intelligence (AI) Trends', *Mad Scientist Laboratory*, 14 December 2017, madsciblog. tradoc.army.mil/11-artificial-intelligence-ai-trends/.
23 Natasha Lomas, 'Lyrebird Is a Voice Mimic for the Fake News Era', *Tech Crunch* (blog), 25 April 2017, techcrunch.com/2017/04/25/lyrebird-is-a-voice-mimic-for-the-fake-news-era/.
24 Lee Rainie and Janna Anderson, 'Code-Dependent: Pros and Cons of the Algorithm Age', *Pew Research Center*, 8 February 2017, www.pewresearch.org/internet/2017/02/08/code-dependent-pros-and-cons-of-the-algorithm-age/.
25 Mick Ryan, *Human–Machine Teaming for Future Ground Forces* (Washington, DC: Centre for Strategic and Budgetary Assessments, 2018), 10.

specialised AI and data analytics and more capable and robust robotic systems. We are already seeing examples of militarised drone swarms, networked unmanned aerial vehicles and surface vessels for maritime surveillance, and a dramatic increase in highly autonomous decision support systems in the commercial sectors.

Information warfare. Information actions are a prerequisite for success.[26] While the importance of winning the narrative has not substantially increased over the last decade, the number of influencing actors has. It is now possible for minor parties or individuals to have a disproportionate effect on the population—both domestically and in the operational environment. The influence of fake news and external parties in recent elections is, at best, a warning of how opinions can be shaped based on misinformation and, at worst, a potential threat to the legitimacy of democratic governments. In a post-truth world, in attempting to control narratives, being right may be less important than being first.

Space. Once a preserve of the superpowers, access to space is now a commercial reality—one being driven by commercial actors such as SpaceX and Blue Origin in the US as well as private sector firms from China—and is highly contested. The number of cube (or small) satellite launches per year has seen a dramatic rise in the last five years, from a dozen to over 250. This number is likely to exponentially increase as the launch vehicles adjust to the demand. Access to space provides states and non-state actors increased surveillance and communication ability at reduced costs. Conversely, competition for access is increasing due to overcrowding (space debris) and the weaponisation of space.[27] For example, the Kessler Syndrome describes a plausible reality where a cascading collision in low orbit creates a high density of objects that effectively denies access to space (and its accompanying functions such as navigation, communication and surveillance). This could result from an accidental or intentional act. It is unlikely that space will remain demilitarised or will return to state control in the near future.

---

26   Head Modernisation and Strategic Planning, *Adaptive Campaigning 09: Army's Future Land Operating Concept* (Canberra: Directorate of Army Research and Analysis, 2009), 50.

27   Joe Pappalardo, 'No Treaty Will Stop Space Weapons', *Popular Mechanics*, 25 January 2018, www.popularmechanics.com/space/satellites/a15884747/no-treaty-will-stop-space-weapons/.

Cybersecurity. As South-East Asian economies digitalise and internet penetration increases in the region, there is a corresponding growth in the regional cybersecurity risk.[28] Differing rates of modernisation present an opportunity for exploitation by transnational cybercrime.[29] Regional measures have focused on capacity building and have not established the legal framework necessary to combat trans-border criminal cyber activities. Failure to develop a collaborative and credible multilateral cyber deterrence mechanism will both exacerbate traditional trans-border criminal activity within the region, such as human trafficking, drug dealing and piracy, as well as compromise the confidentiality and integrity of data, further undermining economic benefit and sovereign security risk.

Climate change and resource scarcity

Climate change, regardless of cause, remains the 'single biggest threat to life, security and prosperity on Earth'.[30] For example, global warming has seen a more rapid melting of the polar ice caps, resulting in a commensurate rise in sea levels and a change in ocean temperature. A change in sea level is likely to both threaten coastal regions and contaminate freshwater sources. Rising sea temperatures, and the corollary rise in acidity levels, are likely to reduce or remove traditional fishing grounds.[31] Combined, these drivers are threatening coastal communities across the globe and increasing the risk to stability. The impact of climate change in our region will continue to be dramatic and obvious. We are more likely to see an increase in intense monsoons and cyclones, coral bleaching and irregular rain patterns that will substantially degrade infrastructure and living conditions. A state's vulnerability to climate change can be described as a function of the impact of physical changes (e.g. rise in sea level or loss of fresh water), sensitivity to climate hazards (e.g. dependence on agriculture) and their capacity to make steps to improve their resilience. The country at greatest risk in Australia's

---

28   L. Hanlen, 'Future Direction for the Digital Economy: Implications for Australia', *Synergy*, presentation as part of Senior Level National Security Course (Canberra: The Australian National University, 21 November 2018).

29   Nicholas Thomas, 'Cyber Security in East Asia: Governing Anarchy', *Asian Security* 5 (2009): 3–23, doi.org/10.1080/14799850802611446.

30   Patricia Espinosa, UN Climate Change Executive Secretary, as quoted in United Nations, *UN Climate Change Annual Report 2017* (Bonn: United Nations, 2018), 5, unfccc.int/sites/default/files/resource/UN ClimateChange_annualreport2017_final.pdf.

31   Our oceans have become more acidic by absorbing and storing the high levels of atmospheric carbon dioxide (CO_2) emitted mainly from human activities: StatsNZ, 'Ocean Acidification', 25 August 2022, stats.govt.nz/indicators/ocean-acidification/.

region is PNG, not the Pacific Islands, due to sensitivity and capacity for change.[32] For instance, Fiji is adjusting its economy to orientate away from agriculture and towards services and tourism.

Climate change appears to be affecting the Asian monsoon, with official studies showing a 4.5 per cent decline in monsoonal rain in the three decades to 2009 and a disruption to normal rain patterns. Yet, at the same time, simulations predict that climate change will result in increased monsoonal precipitation over South Asia, East Asia and the western Pacific Ocean. Increased precipitation, along with increased glacial melting, could potentially have devastating consequences, as witnessed in Pakistan in 2010 when a fifth of the country's total land area was affected by flooding.[33]

While there has not been a significant rise in the number of natural disasters, those that do occur are much more intense. These disasters are costly in terms of infrastructure and human life. Concurrently, rising temperatures are posing significant risks to agricultural systems where monoculture production is threatened and the environment is struggling to adapt. Resources are getting more difficult to secure and states are starting to position themselves for long-term survival at the expense of others.

Resource scarcity — Water, fuel and computing materials

The lack of access to fresh water (referred to as water stress) is increasingly accelerating instability. A recent report stated that water stress is rising with an escalating global population and the impact of climate change.[34] Water stress is more likely to occur in developing areas, such as regions of Africa, the Middle East and South America. Water, while not yet a cause for war,

32   Martina Grecequet, Ian Noble, and Jessica Hellmann, 'Many Small Island Nations Can Adapt to Climate Change with Global Support', *The Conversation*, 16 November 2017, theconversation.com/many-small-island-nations-can-adapt-to-climate-change-with-global-support-86820.
33   Cécile Levacher, 'Climate Change in the Tibetan Plateau Region: Glacial Melt and Future Water Security', *Future Directions International*, 29 May 2014, web.archive.org/web/20210726165659/https://www.futuredirections.org.au/publication/climate-change-in-the-tibetan-plateau-region-glacial-melt-and-future-water-security/.
34   CNA, *The Role of Water Stress in Instability and Conflict*, Report CRM-2017-I-016532 (Arlington: CNA Analysis, 2017), www.cna.org/archive/CNA_Files/pdf/crm-2017-u-016532-final.pdf.

is increasingly seen as a multiplier factor in regional instability. There is potential for state or non-state actors to leverage water to control and coerce civilian populations.

State dependence on petroleum to power their economies and militaries creates a risk. Access to petroleum, and the associated refining, is limited, with many states divesting themselves of this capability in the name of efficiency. As this fuel must be transported, often through highly competitive zones, states are increasingly captive to foreign interference unless alternate fuel sources are found. This century has seen a rapid increase in the development of alternate fuels, however, only 12.1 per cent of global power currently comes from 'clean' sources.[35] Many states have declared an intent to transition to clean or renewable energy, and the world leader is China, which accounts for 45 per cent of global investment in renewable energy. Until alternative fuels are more readily manufactured and accessed, petroleum remains a finite resource and a potential vulnerability for nations that rely on it.

Australia is almost totally reliant upon imports for its aviation fuel, marine diesel, gasoline and motor diesel. This complex and lengthy supply chain is vulnerable to a number of different factors that vary from global fuel prices to terrorism activity. Fortunately, the reality is that Australia's fuel supply is reasonably assured unless there is an outbreak of war in the Indo-Pacific region.[36] Over the longer term, Australia's deposits of lithium may reduce reliance on this supply chain, however, in the near to medium future a large-scale conflict will challenge Australia's supply of fuel and the effectiveness of the land force.

Further, reliance upon mature energy and manufacturing techniques and processes could pose a developmental risk in the future. As battery and microprocessor technology develops, traditional resources for these may change (away from lithium and silicon for example) and shift power and economic advantage to those that control them. Rare-earth minerals (REM) are an example for modern touchscreen technology. China mostly held

---

35 United Nations Environment's Economy Division, Frankfurt School – UNEP Collaborating Centre for Climate and Sustainable Energy Finance and Bloomberg New Energy Finance, *Global Trends in Renewable Energy Investment 2018* (Frankfurt: United Nations Environment Program, Frankfurt School of Finance and Management, 2018), europa.eu/capacity4dev/unep/documents/global-trends-renewable-energy-investment-2018.
36 Keyurkumar Patel, 'Australia's Petroleum Supply and Its Implications for the ADF', *Australian Defence Force Journal* 204 (2018): 71–76.

the largest deposits of REM until Japan developed a technique to harvest them cheaply from the Japanese exclusive economic zone.[37] Resources such as water and fuel are readily identified as finite and vital materials for populations and their military forces, however, REM will be increasingly relied on by nations where technology is ubiquitous to their way of life. As a result, those that control the production and distribution of REM may gradually increase their economic advantage.

'Democratisation of warfare' and decreasing state influence

Relevance of states. Many states are under pressure to maintain relevance and are, consequently, likely to find it increasingly difficult to maintain a monopoly on the use of force. Traditionally relied upon to deliver security and social welfare, states are being challenged by increasing participation of highly skilled individuals, non-government organisations, large multinational firms and subnational actors. Individuals are less likely to be defined by nationality and more by ideology. Over the next two decades, we can expect to see a transition from a state-centric world towards a 'multilevel, multinodal' world, characterised by competition for population influence by institutions, city-states, corporations and individuals.

In this environment, private and non-state groups, in the absence of strong or legitimate states, will increasingly turn to violence to advance their political, social, ideological or economic goals. Substate and transnational actors, enabled by the ability to rapidly share information through mobile devices and associated social media platforms, can encourage collective action and popular movements in hours, rather than months or years. Small, motivated groups or even lone radicalised individuals will continue to wield enormous influence using 'off-grid' mesh networks to disrupt the political and social order of a nation. The rise of non-state-sponsored violence is likely to result from several important trends:[38]

---

37   Frederick Kuo, 'Is Japan's Rare Earth Discovery Fool's Gold?', *The Interpreter*, 18 April 2018, www.lowyinstitute.org/the-interpreter/japan-s-rare-earth-discovery-fool-s-gold.

38   These trends identified in: Joint Chiefs of Staff, *Joint Operating Environment 2035: The Joint Force in a Contested and Disordered World* (Washington, DC: Joint Chiefs of Staff, 14 July 2016), 14.

- **Adaptive irregular/substate adversaries**. Adversaries will continue developing capabilities to avoid or withstand state technological overmatch. Many criminal and terrorist groups are likely to combine relatively cheap, accessible and potentially disruptive technologies—such as social media, smartphones, additive printing, robotic and autonomous systems—to degrade or even challenge state dominance in the future.

- **Disruptive manufacturing technologies and the urban arsenal**. The proliferation of technology and a wide range of manufacturing capabilities in many urban areas will likely continue over the next two decades and might lead to novel advances, including pervasive intelligence, surveillance and reconnaissance instruments and relatively cheap and simple, yet effective strike assets such as 3D-printed drones or sophisticated improvised explosives. It is likely that manufacturing will be distributed throughout the urban battlespace as novel technology enables more people/companies to make locally. This will challenge traditional network analysis of adversary resources when more people can generate what is needed.

- **Weaponisation of commercial technologies**. Over the next two decades, a greater number of people will become connected and begin to take advantage of mobile technologies. Driven largely by expected advances in computerisation, miniaturisation and digitisation, potential adversaries will likely have greater access to more sophisticated weaponry that requires fewer and fewer sophisticated users for effective employment.

Conclusion

A more interconnected world will continue to increase—rather than reduce—differences over ideas and identities. Populism will increase over the next two decades should current demographic, economic, and governance trends hold. So, too, will exclusionary national and religious identities, as the interplay between technology and culture accelerates and people seek meaning and security in the context of rapid and disorienting economic, social, and technological change. Political leaders will find identity politics useful for mobilizing supporters and consolidating political control. Similarly, identity groups will become more influential. Growing access to information and communication tools will enable them to better organize and mobilize [at risk of being tracked and persecuted by existing state security infrastructure]—around political issues, religion, values, economic interests, ethnicity, gender, and lifestyle. The increasingly segregated information and media environment will harden identities.[39]

Technological advances, particularly in the areas of AI, robotics, space and cybersecurity are shaping how societies, and their armed forces, operate and evolve. These advances will continue at speed and interact with developments resulting from regional population growth, the effects of climate change and the competition for finite resources. Complexity, while not a new phenomenon, should be an expected condition of the operating environment. There will be an increased number of capable actors who are likely to have independent agendas that will be difficult to accommodate. States can still be expected to be the dominant actors for the foreseeable future—particularly in the execution of military operations and aggressive statecraft.

References

'Artificial Intelligence (AI) Trends'. *Mad Scientist Laboratory*, 14 December 2017. madsciblog.tradoc.army.mil/11-artificial-intelligence-ai-trends/.

Clausewitz, Carl von. *On War*, edited and translated by Michael Howard and Peter Paret. New York: Knopf, 1976.

CNA. *The Role of Water Stress in Instability and Conflict*. Report CRM-2017-I-016532. Arlington: CNA Analysis, 2017. www.cna.org/archive/CNA_Files/pdf/crm-2017-u-016532-final.pdf.

David, Tal. 'Our Future Is Urban: Future of Cities P1'. *Quantum Run*, 15 September 2020. www.quantumrun.com/Prediction/our-future-urban-future-cities-p1.

Diab, Jasmin, and Nathan K Finney. 'Whither the Hover Tank? Why We Can't Predict the Future of Warfare'. *Modern War Institute*, 6 December 2018. mwi.westpoint.edu/whither-hover-tank-cant-predict-future-warfare/.

Glaser, April, and Rani Molla. 'The Number of Robots Sold in the U.S. Will Jump Nearly 300 Percent in Nine Years'. *ReCode*, 3 April 2017.

---

39 National Intelligence Council, *Global Trends: Paradox of Progress*, Report no. NIC 2017-001 (United States: National Intelligence Council, 9 January 2017), 17.

Grecequet, Martina, Ian Noble, and Jessica Hellmann. 'Many Small Island Nations Can Adapt to Climate Change with Global Support'. *The Conversation*, 16 November 2017. theconversation.com/many-small-island-nations-can-adapt-to-climate-change-with-global-support-86820.

Hanlen, L. 'Future Direction for the Digital Economy: Implications for Australia'. *Synergy*, presentation as part of Senior Level National Security Course. Canberra: The Australian National University, 21 November 2018.

Head Modernisation and Strategic Planning. *Adaptive Campaigning 09: Army's Future Land Operating Concept.* Canberra: Directorate of Army Research and Analysis, 2009.

Joint Chiefs of Staff. *Joint Operating Environment 2035: The Joint Force in a Contested and Disordered World.* Washington, DC: Joint Chiefs of Staff, 14 July 2016.

Kuo, Frederick. 'Is Japan's Rare Earth Discovery Fool's Gold?' *The Interpreter,* 18 April 2018. www.lowyinstitute.org/the-interpreter/japan-s-rare-earth-discovery-fool-s-gold.

Levacher, Cécile. 'Climate Change in the Tibetan Plateau Region: Glacial Melt and Future Water Security'. *Future Directions International*, 29 May 2014. web.archive.org/web/20210726165659/https://www.futuredirections.org.au/publication/climate-change-in-the-tibetan-plateau-region-glacial-melt-and-future-water-security/.

Lomas, Natasha. 'Lyrebird Is a Voice Mimic for the Fake News Era'. *Tech Crunch* (blog), 25 April 2017. techcrunch.com/2017/04/25/lyrebird-is-a-voice-mimic-for-the-fake-news-era/.

Manuel, Anja. 'China Is Quietly Reshaping the World'. *The Atlantic*, 17 October 2017. www.theatlantic.com/international/archive/2017/10/china-belt-and-road/542667/.

Matisek, Jahara, and Ian Bertram. 'The Death of American Conventional Warfare: It's the Political Willpower, Stupid'. *The Strategy Bridge*, 5 November 2017. thestrategybridge.org/the-bridge/2017/11/5/the-death-of-american-conventional-warfare-its-the-political-willpower-stupid.

National Intelligence Council. *Global Trends: Paradox of Progress.* Report no. NIC 2017-001. United States: National Intelligence Council, 9 January 2017.

NATO. *Strategic Foresight Analysis Report 2017.* Norfolk: NATO Strategic Analysis Branch, 2017.

New Zealand Foreign Affairs and Trade. *Pacific Diagnostic 2015–2035.* Wellington: NZ MFAT, 2015.

Pappalardo, Joe. 'No Treaty Will Stop Space Weapons'. *Popular Mechanics*, 25 January 2018. www.popularmechanics.com/space/satellites/a15884747/no-treaty-will-stop-space-weapons/.

Patel, Keyurkumar. 'Australia's Petroleum Supply and Its Implications for the ADF'. *Australian Defence Force Journal* 204 (2018): 71–76.

Peng, Xiujian. 'China's Population Is Now Inexorably Shrinking, Bringing Forward the Day the Planet's Population Turns down'. *The Conversation*, 19 January 2023. theconversation.com/chinas-population-is-now-inexorably-shrinking-bringing-forward-the-day-the-planets-population-turns-down-198061.

Rainie, Lee, and Janna Anderson. 'Code-Dependent: Pros and Cons of the Algorithm Age'. *Pew Research Center*, 8 February 2017. www.pewresearch.org/internet/2017/02/08/code-dependent-pros-and-cons-of-the-algorithm-age/.

Ryan, Mick. *Human–Machine Teaming for Future Ground Forces.* Washington, DC: Centre for Strategic and Budgetary Assessments, 2018.

Schwarzer, Daniela. 'Europe, the End of the West and Global Power Shifts'. *Global Policy* 8, no. S4 (2017): 18–26. onlinelibrary.wiley.com/doi/full/10.1111/1758-5899.12437.

Singer, PW. *Wired for War.* Boston: Houghton Mifflin Harcourt, 2009.

StatsNZ. 'Ocean Acidification', 25 August 2022. stats.govt.nz/indicators/ocean-acidification/.

Steinbock, Dan. 'The Global Economic Balance of Power Is Shifting'. *World Economic Forum*, 20 September 2017. www.weforum.org/agenda/2017/09/the-global-economic-balance-of-power-is-shifting.

Thomas, Nicholas. 'Cyber Security in East Asia: Governing Anarchy'. *Asian Security* 5 (2009): 3–23. doi.org/10.1080/14799850802611446.

United Kingdom Ministry of Defence. *Global Strategic Trends: Out to 2045.* London: Ministry of Defence, 2018.

United Nations. *UN Climate Change Annual Report 2017.* Bonn: United Nations, 2018. unfccc.int/sites/default/files/resource/UNClimateChange_annualreport 2017_final.pdf.

United Nations. *World Urbanization Prospects: The 2018 Revision.* New York: United Nations, 2019. population.un.org/wup/.

United Nations Environment's Economy Division, Frankfurt School – UNEP Collaborating Centre for Climate and Sustainable Energy Finance and Bloomberg New Energy Finance. *Global Trends in Renewable Energy Investment 2018*. Frankfurt: United Nations Environment Program, Frankfurt School of Finance and Management, 2018. europa.eu/capacity4dev/unep/documents/global-trends-renewable-energy-investment-2018.

United Nations Population Fund. *Population and Development Profiles: Pacific Island Countries*. Suva: UNFPA, 2014, pacific.unfpa.org/sites/default/files/pub-pdf/web__140414_UNFPAPopulationandDevelopmentProfiles-PacificSub-Region Extendedv1LRv2_0.pdf.

US Army TRADOC G2 Mad Scientist Initiative. 'An Advanced Engagement Battlespace: Tactical, Operational and Strategic Implications for the Future Operating Environment'. *Small Wars Journal*, 24 October 2017. archive.small warsjournal.com/jrnl/art/advanced-engagement-battlespace-tactical-operational-and-strategic-implications-future.

2

The Democratisation of Technology: Opportunities and Threats

Nadya T Bliss

Introduction[1]

The origins of much of today's technology can be traced back to research conducted on behalf of the US Government. The internet, global positioning systems (GPS), microprocessors—the US Government played a role in developing all these technologies, often in pursuit of national security interests. In the process, the initial research supported by the US Government helped spawn entire industries.[2]

In the decades it took for a modest computer network developed by the Department of Defense to take hold and inspire the creation of the internet, the driving force of technological innovation shifted from government to the private sector, from Washington, DC to Silicon Valley. By the time the internet was ubiquitous in the 2000s, industry was the primary driver of new technology, while the US Government's role in technological research and development (R&D) had diminished.

---

1   Editors' note: as mentioned in the Introduction, this chapter was written in 2020, and global events since then may have overtaken some aspects. As the pace of technological and cultural change has continued to accelerate, the editors have opted to present this text as drafted to minimise further delays in publication.
2   National Research Council, *Continuing Innovation in Information Technology* (Washington, DC: The National Academies Press, 2012), 5.

This transition has significantly impacted national security. The US and its allies can no longer overwhelm adversaries through technological superiority. To be sure, the US retains technological pre-eminence over its competitors in many areas, but the advantage that supremacy yields has deteriorated. At the same time, adversaries have access to more of the same technologies than ever before and can use them to their benefit. Foreign meddling in domestic politics is a prime example. It is not new, but the prevalence of social media and the expansion of internet access has resulted in impact at a scale not previously seen, and with minimal risk to the transgressor.

Widespread access to technology has also weakened the resource advantage democracies enjoyed for most of the last half of the 20th century. Adversaries with fewer resources than their larger opponents have always sought to level the playing field by other means. Today, that often takes the form of using technologies developed by American companies against the US and its allies. Again, foreign interference in the 2016 US presidential election is a useful example. Social media companies like Facebook and Twitter (now known as X) were used as weapons to manipulate domestic politics and exacerbate societal divisions,[3] and were simply not prepared or properly incentivised to combat such efforts. They still are not.

The democratisation of technology—meaning the rapid dispersion of new technologies that are designed and created to be used by as many people as possible—has had some tremendously positive effects on society. But there have also been negative consequences, primarily in terms of national security. Democracies, with their emphasis on freedom of speech, competitive markets and rule of law, are particularly vulnerable to these threats and must band together and invest in new capabilities that prioritise national security interests over market share.

This will require a renewed dedication to technological R&D on the part of governments, as well as a willingness to break free from old R&D funding habits. Governments must incentivise interdisciplinary research in a way they have not before, to spur the type of research that explores problems from more than the technological standpoint. In short, democratic governments must reclaim their role as key drivers of innovation.

---

3   Robert S Mueller, *Report on the Investigation into Russian Interference in the 2016 Presidential Election* (Washington, DC: US Department of Justice, 2019), 14–15, upload.wikimedia.org/wikipedia/commons/e/ e4/Report_On_The_Investigation_Into_Russian_Interference_In_The_2016_Presidential_Election.pdf.

One current opportunity area to do so is the challenge of combating the spread of disinformation for political purposes. It is a grand challenge facing many democracies, but governments thus far have played a largely hands-off role in curtailing such efforts. Instead, it has fallen to the same private companies that serve as vehicles for the spread of disinformation to combat it. The US Government should join with its allies in the North Atlantic Treaty Organization (NATO) and the Five Eyes partnership to develop a comprehensive framework for addressing the issue and use this effort to re-establish their role as drivers of important security-related research.

Government as the driver of innovation: 1960s to 1970s

A wave of scientific advancement started with a 98-minute orbit around the Earth by a hunk of metal not much larger than a beach ball. Sputnik, the world's first artificial satellite, was launched on 4 October 1957, ushering in the space age and a race for new scientific breakthroughs between the world's two pre-eminent powers.

Taken by surprise and determined to outdo its Cold War adversary, the US responded by investing in scientific research and education. Within a year, the US had created the Advanced Research Projects Agency (ARPA) to conduct research on military technology, and the National Aeronautics and Space Administration (NASA) to advance scientific knowledge. Both organisations would play pivotal roles in technological advancements to come. The US also responded to the challenge by greatly expanding federal funding for science, technology, engineering and mathematics (STEM) education, and by creating the first federal student loan program to encourage more people to attend college.

This full-throated response to a complex, immediate threat helped shape the world we have today and established the US Government as a main driver of technological advancement. The most well-known example is the creation of ARPANET, the computer network created by ARPA in the late 1960s that laid the foundation for today's internet. Critically, ARPANET was created not by one organisation or person, but by 'an interdependent, collaborative network of academic, industry, and government experts and experimenters'.[4] The government's role was that of driver, helping to identify an immediate application for the new technology and spurring its development in collaboration with other actors. The government continued to play this role as the internet gained steam, and it was a collection of government networks that were aggregated to form the internet.[5]

The impact of the US response to Sputnik can be seen beyond the internet, in the growth of the US economy and, indeed, in our very culture. Nearly every aspect of information technology that we rely on today bears the stamp of US Government support at some point in its development.

This is the period that solidified the foundation of the US innovation ecosystem, a mutually beneficial relationship among the federal government, universities and the public sector to advance new technologies, from basic research to private sector application to consumers. This partnership between federal R&D and the private sector has proven to be an effective model. Often, R&D that is critical to the nation does not carry a strong enough financial incentive to warrant investment by the private sector, so the government steps in and fills a crucial gap.[6]

ARPA is a useful example. Created with the mission of making the pivotal early technology investments that create or prevent strategic surprise for US national security, ARPA was designed to pursue opportunities for transformational change rather than incremental advances.[7] While its research goals were motivated by national security concerns, it routinely produced basic research that led to significant economic benefits for the nation.[8]

---

4   National Academies of Sciences, Engineering, and Medicine, *Continuing Innovation in Information Technology: Workshop Report* (Washington, DC: The National Academies Press, 2016), www.nap.edu/read/23393/chapter/1#iv.
5   National Academies of Sciences, Engineering, and Medicine, *Continuing Innovation in Information Technology*.
6   Wilbur L Ross Jr and Walter Copan, *Return on Investment Initiative for Unleashing American Innovation*, National Institute of Standards and Technology Special Publication 1234 (Washington, DC: National Institute of Standards and Technology, April 2019), 13, nvlpubs.nist.gov/nistpubs/Special Publications/NIST.SP.1234.pdf.
7   DARPA, *Creating Breakthrough Technologies for National Security*, Distribution Statement A (Washington, DC: DARPA, July 2017).
8   Sean Pool and Jennifer Erickson, *The High Return on Investment for Publicly Funded Research* (Washington, DC: The Center for American Progress, 2012), www.americanprogress.org/issues/economy/reports/2012/12/10/47481/the-high-return-on-investment-for-publicly-funded-research/.

In fact, the US innovation ecosystem helped spur the development of some of today's leading business sectors, including multibillion-dollar industries such as broadband and mobile, cloud computing, personal computing and microprocessors. These business sectors often were not profitable until decades after basic research on the topic was conducted.[9]

Government takes a backseat: 2000 to today

Ironically, the growth of the internet and the business sectors that have developed around it have led to a reduction in the US Government's role in the R&D of new technologies. This is partly due to growing industry investments in R&D, and partly to stagnant public investment. According to data from the Organisation for Economic Co-operation and Development, US funding for R&D as measured by share of gross domestic product has declined by over 15 per cent since 1995.[10]

The reduction of the government's role in R&D has impacted national security. Industry's motivations in R&D are different than government's: their R&D tends to be shorter term and focused on applied research, with the goal of disbursing technology rapidly and widely with little consideration for how adversaries may abuse that technology.

While the government remains the driver of technological innovation in hard power—force projection, nuclear forces, missile defence—it is one customer among many or a bystander on some critical issues facing the nation. These include infrastructure resilience and biodefense. Worse, the government is the last to know in areas where adversaries can use existing technologies to exploit bedrock democratic principles such as freedom of speech (Figure 2.1). This is why the spread of disinformation online caught so many in the US national security community by surprise. Another area of vulnerability is the potential exploitation of the algorithms that drive artificial intelligence decision-making. The US Government must renew its commitment to R&D if it is to effectively combat the challenges stemming from the abuse of today's ubiquitous technologies.

---

9   National Research Council, *Continuing Innovation in Information Technology*, 5.
10   Task Force on American Innovation, *Second Place America? Increasing Challenges to U.S. Scientific Leadership* (United States: Task Force on American Innovation, May 2019), 12.

U.S. Government's role in IT innovation and challenges

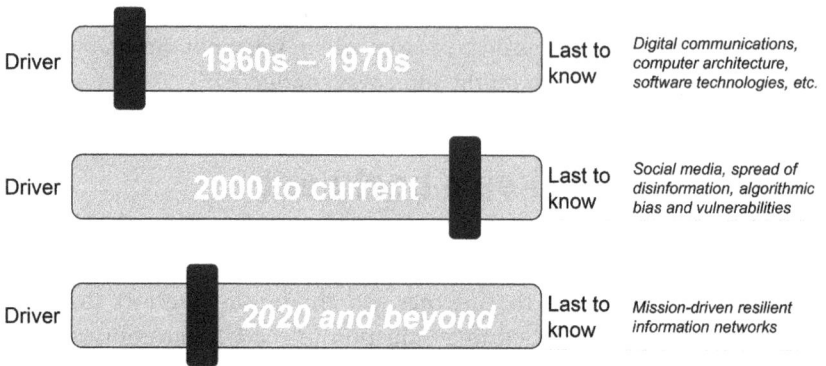

Driver | 1960s – 1970s | Last to know — Digital communications, computer architecture, software technologies, etc.

Driver | 2000 to current | Last to know — Social media, spread of disinformation, algorithmic bias and vulnerabilities

Driver | 2020 and beyond | Last to know — Mission-driven resilient information networks

Figure 2.1: US Government role in development of IT innovations.
Source: Nadya T Bliss, from presentation 'The Democratization of Technology: Opportunities and Threats'.

A new Sputnik moment

In 2014, the Russian Internet Research Agency (IRA) sent employees to the US on an intelligence-gathering mission, according to a report on election interference in the 2016 presidential election. Over the next two years, the IRA would conduct a sophisticated 'information warfare' campaign using social media as its primary weapon. By the end of the 2016 US election, the IRA had the ability to reach millions of US citizens through their social media accounts, including multiple US political figures who retweeted IRA-created content. Twitter notified approximately 1.4 million people it believed may have been in contact with an IRA-controlled account.[11] Facebook's CEO estimates the IRA reached as many as 126 million people through its Facebook accounts, and that the IRA spent approximately $100,000 on more than 3,000 ads on Facebook and Instagram.[12]

---

11   Mueller, *Report on the Investigation into Russian Interference.*
12   *Hearing before the United States Senate Committee on the Judiciary and the United States Senate Committee on Commerce, Science and Transportation*, 10 April 2018 (testimony of Mark Zuckerberg, chairman and CEO of Facebook), www.judiciary.senate.gov/imo/media/doc/04-10-18%20Zuckerberg%20Testimony.pdf.

Efforts by geopolitical competitors to influence elections and political debates in other nations are hardly new. Their scale and potential impact have, however, been amplified because of new technologies. The Russian effort to influence the 2016 American presidential election hinged on the use of social media to rapidly spread misleading or false information and heighten political tensions.

The US Government is not alone in being targeted by such efforts. According to a report from the National Endowment for Democracy (NED), authoritarian governments are increasingly using disinformation to influence elections beyond their borders. Russian efforts in this area include French, German and American elections in 2016 and 2017, the 2018 Czech presidential election and the 2017 vote on Catalonian secession from Spain. As the NED report notes:

> In each of these cases, Moscow used a combination of state-owned international news outlets, smaller news sites linked to Moscow, and automated social media accounts, sometimes in tandem with leaks of stolen documents and communications.[13]

Despite the threat this poses to the basic tenets of democracy, the US Government and its allies have yet to develop a robust response or a framework for combating the issue. There is limited investment in new technologies or research on the root causes of disinformation and why it is effective, and thus far insufficient appetite to impose new regulations on the primary vehicles for the spread of disinformation, namely, social media companies. Companies' efforts to curb the problem by building new tools to increase transparency and identify disinformation are helpful, but not sufficient. The root causes of the problem run deeper and are baked into the financial incentive structure of the platforms, or as Dipayan Ghosh and Ben Scott of New America call it, 'the political economy of digital information markets'.[14]

This is a challenge that requires a governmental response. The severity of the threat, the complexity of the problem and the incentive structure to address the issue demand a coordinated, strategic, long-term effort built around a renewed R&D program.

---

13   Dean Jackson, *Issue Brief: How Disinformation Impacts Politics and Publics* (Washington, DC: National Endowment for Democracy, May 2018).
14   Dipayan Ghosh and Ben Scott, *#DigitalDeceit: The Technologies behind Precision Propaganda on the Internet* (Washington DC: New America, January 2018), 3, www.newamerica.org/public-interest-technology/policy-papers/digitaldeceit/.

This type of response fits within the framework of mission-oriented policies, defined by Professor Mariana Mazzucato as:

> systemic public policies that draw on frontier knowledge to attain specific goals, or 'big science deployed to meet big problems'. The archetypical historical mission is NASA putting a man on the moon.[15]

Mazzucato explains that, under this framework, the role of state organisations is to coordinate and provide direction to private actors when formulating and implementing policies that address societal challenges through innovation.[16]

While this approach echoes the US response to Sputnik and the early days of ARPA and NASA, it is different and broader in scope in fundamental ways, according to Mazzucato:

> While missions of the past aimed to develop a particular technology (with the achievement of the technological objective signalling that the mission was accomplished), contemporary missions address broader and more persistent challenges, which require long-term commitments to the development of technological solutions.[17]

A response commensurate with the problem

This long-term commitment should include not only increased governmental funding for R&D, but also a focus on interdisciplinary research in order to understand the various underlying causes and potential solutions, significant international cooperation, increased resources for public education on new technologies, and the creation of a formal process for exploring potential abuses of new technology that is conducted simultaneously with the development of said technologies.

---

15   Mariana Mazzucato, *Mission-Oriented Innovation Policy: Challenges and Opportunities*, Working Paper IIPP WP 2017-01 (London: UCL Institute for Innovation and Public Purpose, 2017), 7, www. ucl.ac.uk/bartlett/public-purpose/sites/public-purpose/files/moip-challenges-and-opportunities-working-paper-2017-1.pdf.
16   Mazzucato, *Mission-Oriented Innovation Policy*, 4.
17   Mariana Mazzucato, *A Mission-Oriented Approach to Building the Entrepreneurial State* (Swindon: Innovate UK, November 2014), 12, web.archive.org/web/20201218105633/https://marianamazzucato. com/wp-content/uploads/2014/11/MAZZUCATO-INNOVATE-UK.pdf.

- *Increased governmental funding for R&D*: In 1960, the US Government spent twice as much as industry on R&D.[18] Today, those figures are reversed, and geopolitical competitors like China are catching up to US R&D funding in real dollars. To reclaim their role as important drivers of technological innovation, governments must dedicate more resources to both basic and applied R&D.

- *Focus on interdisciplinary research*: Current R&D funding structures typically disincentivise interdisciplinary research. Metrics for success are often based on technological requirements, relegating many research projects to the engineering domain. This tendency leads to lost opportunities to explore problems from more than one angle and to develop more holistic, longer-term solutions. Funding agencies should prioritise interdisciplinary research and the development of meaningful metrics to measure progress on interdisciplinary research.

- *Significant international cooperation*: The specific challenge of disinformation is a global problem that capitalises on globally available technology. As such, it requires a global response. The world's major democratic security alliances, NATO and the Five Eyes intelligence partnership, should prioritise responding to the threat disinformation poses to member countries and develop comprehensive frameworks aimed at stemming its influence.

- *Increased resources for public education on new technologies*: Disinformation campaigns have capitalised on recent innovations in technology and the interconnectedness of our world, but at root they are successful because of human failures to recognise falsehoods when we see them. A sustained public education effort aimed at improving societal technological literacy could help address this issue and others. This will only gain in importance as more decisions are made with the aid of automated systems. It is important that people understand how those decisions are being made.

- *A formal process for exploring potential abuses of new technologies*: R&D funders should prioritise technological projects that leverage the critical inquiry principles found in social sciences to complement the work of engineers and technologists. Structuring programs to better understand and explore potential abuses provides a useful entryway into conducting more interdisciplinary research, and could be listed as a requirement

---

18   Congressional Research Service, *U.S. R&D Funding and Performance: Fact Sheet*, Report R44307 Version 10 (Washington, DC: Congressional Research Service, 29 June 2018).

by R&D funders. Conducting this work up-front can help developers account for and potentially prevent some potential abuses when creating the new technology.

Addressing the threat of disinformation also requires a framework that better defines the problem and its various manifestations, considers possible R&D elements and assigns the domains that could help drive responses to the various forms of the challenge. The appendix to this chapter includes such a framework.

Conclusion

The reduction of the role of the US Government, and the concurrent role of the private sector in funding research and developing new technologies, has resulted in an erosion of the technological lead that the US had previously enjoyed for most of the last half of the 20th century. Adversaries with fewer resources have access to more of the same technologies than ever before and can use them to their benefit. Thus, while the democratisation of technology—that is, the rapid dispersion of new technologies that are designed and created to be used by as many people as possible—has had some tremendously positive effects on society, it has also delivered negative consequences, primarily in terms of national security.

Democracies that have driven these technologies also have inherent vulnerabilities resulting from the key characteristics of their societies—the emphasis on freedom of speech, competitive markets and the rule of law. These have eclipsed national security interests in many areas, and the latter now requires the collective attention of democratic states.

During the early stages of the Cold War, the US responded to the 'Sputnik moment' by galvanising its scientific research and education programs, and the US Government became a driver of technological advancement. There has since been a tilting of investment from the public sector to the private, leaving the US Government less directly involved with technological change. Challenges today are also more nuanced and potentially come from a variety of actors. As this chapter has noted, the US is facing direct interference in a number of democratic processes such as elections, directly instigated by Russia and its IRA via social media. The combination of the diffusion of actors and technology, and the reduction if the role of the US Government as a driver of change, have left it and other Western states exposed in terms of national security.

The US and other Western governments are now poorly placed to identify where adversaries can use existing technologies to exploit fundamental democratic principles such as freedom of speech, while infrastructure resilience and biodefense are further areas of weakness. Therefore governments must revise their approach to scientific and technology R&D to address today's security challenges, with information manipulation a prime example of a problem in need of a different approach. The rapid disbursement of new technologies has significantly expanded the impact of disinformation campaigns, and they are now a serious security threat that requires international cooperation and a sustained, strategic response.

But this challenge—along with the international competition to become the leader in next-generation artificial intelligence, 5G and quantum computing—also offers opportunities: to renew American leadership in technological R&D, to centre that leadership around solving societal challenges, and to promote forward-looking R&D that leverages the strengths of multiple disciplines and seeks to proactively identify and head off potential problems around new technology. Doing so could help address immediate problems, and lead to a new age of innovation and discovery.

References

Congressional Research Service. *U.S. R&D Funding and Performance: Fact Sheet.* Report R44307 Version 10. Washington, DC: Congressional Research Service, 29 June 2018.

DARPA. *Creating Breakthrough Technologies for National Security.* Distribution Statement A. Washington, DC: DARPA, July 2017. www.darpa.mil/attachments/DARPA_Fact_Sheet_1_07-25-17.pdf.

Ghosh, Dipayan, and Ben Scott. *#DigitalDeceit: The Technologies behind Precision Propaganda on the Internet.* Washington, DC: New America, January 2018. www.newamerica.org/public-interest-technology/policy-papers/digitaldeceit/.

Hearing before the United States Senate Committee on the Judiciary and the United States Senate Committee on Commerce, Science and Transportation, 10 April 2018 (testimony of Mark Zuckerberg, chairman and CEO of Facebook). www.judiciary.senate.gov/imo/media/doc/04-10-18%20Zuckerberg%20Testimony.pdf.

Jackson, Dean. *Issue Brief: How Disinformation Impacts Politics and Publics.* Washington, DC: National Endowment for Democracy, May 2018.

Mazzucato, Mariana. *A Mission-Oriented Approach to Building the Entrepreneurial State.* Swindon: Innovate UK, November 2014. web.archive.org/web/2020 1218105633/https://marianamazzucato.com/wp-content/uploads/2014/11/ MAZZUCATO-INNOVATE-UK.pdf.

Mazzucato, Mariana. *Mission-Oriented Innovation Policy: Challenges and Opportunities.* Working Paper IIPP WP 2017-01. London: UCL Institute for Innovation and Public Purpose, 2017. www.ucl.ac.uk/bartlett/public-purpose/ sites/public-purpose/files/moip-challenges-and-opportunities-working-paper-2017-1.pdf.

Mueller, Robert S. *Report on the Investigation into Russian Interference in the 2016 Presidential Election.* Washington, DC: US Department of Justice, 2019. upload. wikimedia.org/wikipedia/commons/e/e4/Report_On_The_Investigation_Into_ Russian_Interference_In_The_2016_Presidential_Election.pdf.

National Academies of Sciences, Engineering, and Medicine. *Continuing Innovation in Information Technology: Workshop Report.* Washington, DC: The National Academies Press, 2016. www.nap.edu/read/23393/chapter/1#iv.

National Research Council. *Continuing Innovation in Information Technology.* Washington, DC: The National Academies Press, 2012.

Pool, Sean, and Jennifer Erickson. *The High Return on Investment for Publicly Funded Research.* Washington, DC: The Center for American Progress, 2012. www.americanprogress.org/issues/economy/reports/2012/12/10/47481/the-high-return-on-investment-for-publicly-funded-research/.

Ross, Wilbur L, Jr, and Walter Copan. *Return on Investment Initiative for Unleashing American Innovation.* National Institute of Standards and Technology Special Publication 1234. Washington, DC: National Institute of Standards and Technology, April 2019. nvlpubs.nist.gov/nistpubs/SpecialPublications/NIST. SP.1234.pdf.

Task Force on American Innovation. *Second Place America? Increasing Challenges to U.S. Scientific Leadership.* United States: Task Force on American Innovation, May 2019.

Appendix: A framework for addressing information manipulation

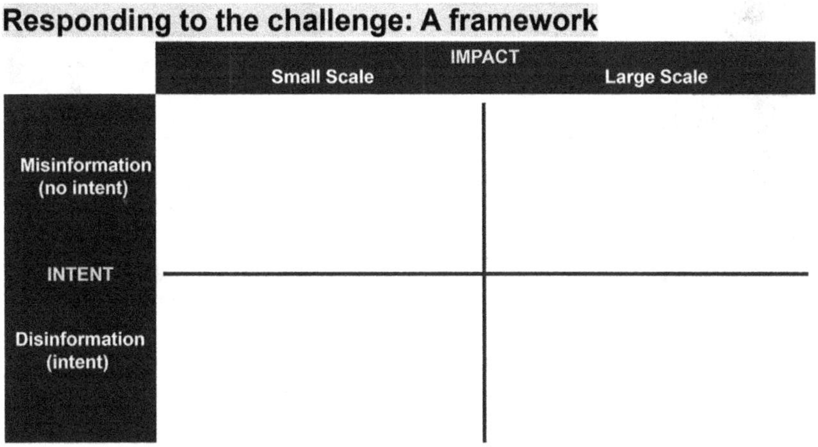

Responding to the challenge: A framework

	IMPACT	
	Small Scale	Large Scale
Misinformation (no intent)		
INTENT		
Disinformation (intent)		

Figure 2.A1: Responding to the challenge: A framework.

Source: Nadya T Bliss, from presentation 'The Democratization of Technology: Opportunities and Threats'.

Responding to the challenge: Definitions and examples

	IMPACT	
	Small Scale	Large Scale
Misinformation (no intent)	o Localized o Human error	o Threatens public safety o Anti-vaccine movement o False emergency warnings
INTENT		
Disinformation (intent)	o Localized o In early stages or with a limited purpose	o Information operations o Foreign interference in elections/domestic politics

Disinformation and misinformation are not new, but current technologies have exacerbated the speed with which they can spread and the scope of their potential impact

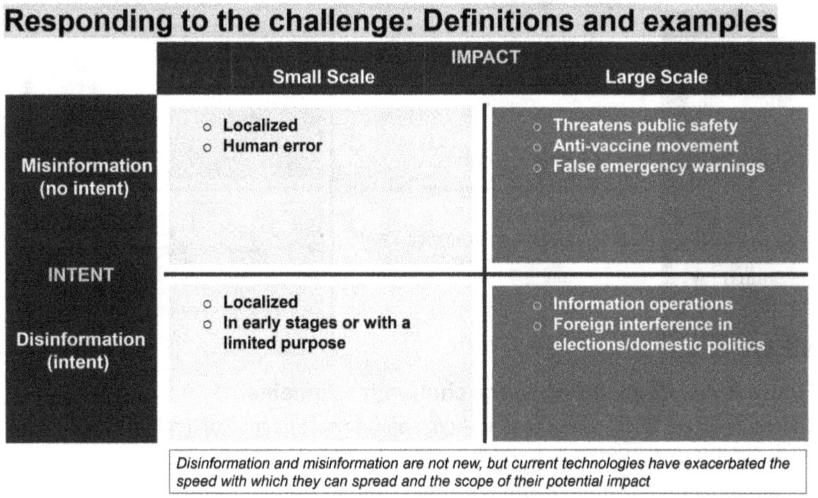

Figure 2.A2: Responding to the challenge: Definitions and examples.

Source: Nadya T Bliss, from presentation 'The Democratization of Technology: Opportunities and Threats'.

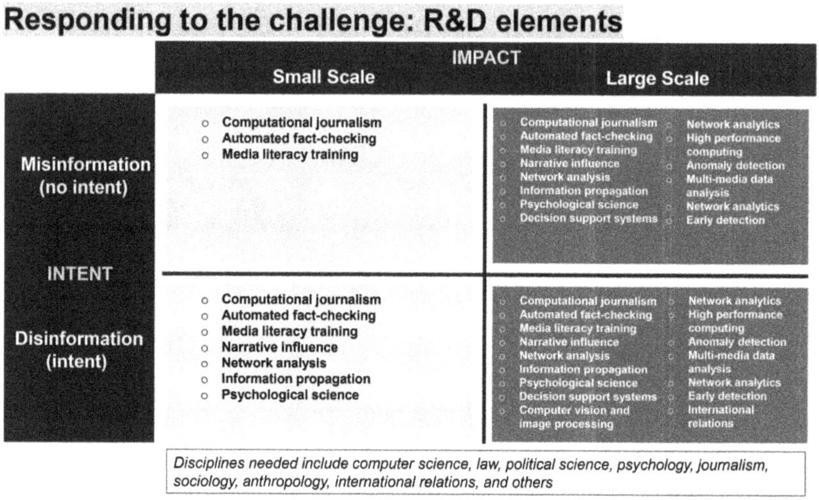

Responding to the challenge: R&D elements

	IMPACT	
	Small Scale	**Large Scale**
Misinformation (no intent)	○ Computational journalism ○ Automated fact-checking ○ Media literacy training	○ Computational journalism ○ Network analytics ○ Automated fact-checking ○ High performance ○ Media literacy training computing ○ Narrative influence ○ Anomaly detection ○ Network analysis ○ Multi-media data ○ Information propagation analysis ○ Psychological science ○ Network analytics ○ Decision support systems ○ Early detection
INTENT		
Disinformation (intent)	○ Computational journalism ○ Automated fact-checking ○ Media literacy training ○ Narrative influence ○ Network analysis ○ Information propagation ○ Psychological science	○ Computational journalism ○ Network analytics ○ Automated fact-checking ○ High performance ○ Media literacy training computing ○ Narrative influence ○ Anomaly detection ○ Network analysis ○ Multi-media data ○ Information propagation analysis ○ Psychological science ○ Network analytics ○ Decision support systems ○ Early detection ○ Computer vision and ○ International image processing relations

Disciplines needed include computer science, law, political science, psychology, journalism, sociology, anthropology, international relations, and others

Figure 2.A3: Responding to the challenge: R&D elements.

Source: Nadya T Bliss, from presentation 'The Democratization of Technology: Opportunities and Threats'.

Responding to the challenge: Domains

	IMPACT	
	Small Scale	**Large Scale**

Misinformation (no intent)	• Community
	  ○ Social networks
	  ○ Industry
	  ○ journalists
		• Governments
		  ○ Departments of Home Affairs / Homeland Security

INTENT		

Disinformation (intent)	• Governments
	  ○ Intelligence agencies and law enforcement
		• International Alliances
		  ○ Five Eyes Intelligence Alliance
		  ○ North Atlantic Treaty Organization

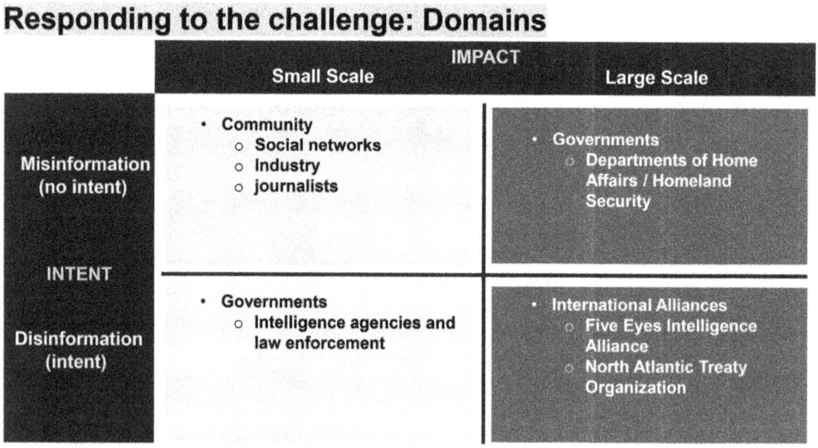

Figure 2.A4: Responding to the challenge: Domains.

Source: Nadya T Bliss, from presentation 'The Democratization of Technology: Opportunities and Threats'.

3

Future Warfare: Developing a Viable Strategy

David Kilcullen

Introduction[1]

Analysing the future of war is a complicated business. Analysts need to account for hundreds of variables across the political, military, economic, social, infrastructure, informational and physical environments and then model these variables along a shifting timescale. They have to identify and, ideally, track changes in system-states within a complex adaptive system, across multiple government and non-state actors.[2] They need to consider internal circumstances, external interactions among allies, adversaries and enemies, the operating environment (across multiple domains) within which interactions occur, the processes and timelines through which we generate warfighting capabilities, and the technologies, platforms, tactics and ethics we employ. This can seem dauntingly complex, especially when the consequences of getting it wrong can be catastrophic.

---

1   Editors' note: as mentioned in the Introduction, this chapter was written in 2020, and global events since then may have overtaken some aspects. As the pace of technological and cultural change has continued to accelerate, the editors have opted to present this text as drafted to minimise further delays in publication.

2   For a critique of this framing, see Brian M Ducote, 'Challenging the Application of PMESII-PT in a Complex Environment' (student monograph, School of Advanced Military Studies, Fort Leavenworth, Kansas, 2010), apps.dtic.mil/sti/pdfs/ADA523040.pdf.

But despite their huge number, these factors fall into three broad categories: enduring features (*constants*) of the environment; continuities (*trends*); and *discontinuities*, sometimes called shocks or strategic surprises. By far the most important of these—and the hardest to account for, in any predictive analysis of future conflict—are discontinuities.

Constants, trends and shocks

Constants either do not change, or change so slowly that analysts often consider them as fixed features of the environment. In Australia's case, a series of influential strategic assessments over many decades—from the Chiefs of Staff Committee's *Appreciation of the Strategical Position of Australia* in February 1946, to Paul Dibb's 1986 *Review of Australia's Defence Capabilities*, to Hugh White's *How to Defend Australia* of 2019—have given extensive consideration to enduring features of our environment.[3] These include Australia's strategic and economic geography: our small population relative to territorial size, our maritime surroundings and related economic reliance on offshore resources and international trade, and our links to European and North American allies who share our civilisational values but reside on the other side of the planet. More broadly, the tyranny of distance is a constant in Australian strategy: the continent's population and infrastructure are clustered in the south-east, so that even operations entirely within Australian territory become expeditionary undertakings, requiring forces to travel thousands of kilometres and then sustain themselves in remote, unsupported areas.[4] The quality of analysis of these constants has varied over time, but their static nature means they can be factored into any assessment relatively easily, giving them utility for understanding the outline parameters—the boundaries of the possible—for future war.

---

3   Chiefs of Staff Committee, 'An Appreciation of the Strategical Position of Australia', in *A History of Australian Strategic Policy since 1945*, ed. Stephan Frühling (Canberra: Department of Defence, 2009), 53–98; Paul Dibb, *Review of Australia's Defence Capabilities* (Canberra: Australian Government Publishing Service, 1986), www.aspistrategist.org.au/wp-content/uploads/2022/02/Review-of-Australias-Defence-Capabilities-1986.pdf; Hugh White, *How to Defend Australia* (Melbourne: Latrobe University Press, 2019).

4   See Robert Stevenson, 'The Tyranny of Distance: Geo-strategy and the New Guinea Campaign of 1914' and Patrick Porter, 'The World Is Not Flat: War and Distance in the Twenty First Century' in *Geo-strategy and War: Enduring Lessons for the Australian Army: The 2015 Chief of Army History Conference*, ed. Peter Dennis (Canberra: Big Sky Publishing, 2015), web.archive.org/web/20200403225438/https://www.army.gov.au/sites/default/files/2015_geo_strategy_and_war.pdf?acsf_files_redirect.

Within these boundaries, *trends* are changes that play out, often in a linear and therefore predictable manner, over a timescale of years or decades. These include demographic, climatic, cultural, technological and economic changes, shifts in the balance of power among competing geopolitical or military actors, and patterns of urban and coastal settlement, among other things. Some are amenable to straight-line projections based on observable changes from a known baseline. The classical example is demography, where for a given start-state (a current population pyramid derived from a government census) with known inputs such as birth and death rates, migration data, infant mortality and life expectancy, analysts can make well-informed judgements as to the size and composition of a given population at any future date.

Other trends (such as new technologies or operational methods) are nonlinear and rely on less verifiable inputs so that their inclusion in any future warfare analysis is a form of guesswork requiring constant monitoring and re-validation. In the case of the Australian Army, *Complex Warfighting* (2005) identified trends of complexity, diversity, diffusion and lethality within the conflict environment, while *Adaptive Campaigning* (2009) identified globalisation, US conventional military primacy, urbanisation, population growth and technological advances as key trends.[5] In *Out of the Mountains* (2013), I further identified population growth, urbanisation, littoralisation and exploding electronic connectivity as mega-trends shaping the conflict environment.[6]

Finally, *discontinuities* are shock events that do not fit within predicted trends, and therefore cannot be derived through linear projections from current conditions. Relatively modest discontinuities may involve rapid shifts in policy or changes in the direction or degree of trends. At their most extreme, however, discontinuities lie outside the previously understood boundaries of the possible, and therefore push these conceptual boundaries outward: in Thomas Kuhn's terms, they are anomalies that trigger crises, thereby forcing a complete rethink (or paradigm shift) about the character of conflict itself.[7]

---

5   Head Modernisation and Strategic Planning, *Adaptive Campaigning 09: Army's Future Land Operating Concept* (Canberra: Directorate of Army Research and Analysis, 2009), 7–13, researchcentre. army.gov.au/sites/default/files/acfloc_2012_main.pdf; David Kilcullen, *Complex Warfighting* (Canberra: Australian Army, 2005), indianstrategicknowledgeonline.com/web/complex_warfighting.pdf.

6   David Kilcullen, *Out of the Mountains: The Coming Age of the Urban Guerrilla* (New York: Oxford University Press, 2013).

7   Thomas S Kuhn, *The Structure of Scientific Revolutions*, 50th ed. (1962, repr. Chicago: The University of Chicago Press, Kindle Edition, 2012), 52–55, 77–80.

Discontinuities of this kind represent 'Black Swans' which, according to Nassim Nicolas Taleb, have three key attributes. First, a Black Swan:

> is an outlier, as it lies outside the realm of regular expectations, because nothing in the past can convincingly point to its possibility. Second, it carries an extreme 'impact'. Third, in spite of its outlier status, human nature makes us concoct explanations for its occurrence after the fact, making it explainable and predictable. I stop and summarize the triplet: rarity, extreme 'impact', and retrospective (though not prospective) predictability.[8]

Taleb's examples include the 1987 stock market crash, the rise of Hitler, the Second World War, the collapse of communism in 1991, the 9/11 terrorist attacks and the 2008 global financial crisis. Pandemics, revolutions and major natural disasters would also fit the bill.

Most future warfare analyses focus on constants and trends, ignoring discontinuities. This is unavoidable to some extent, since by definition we do not know what discontinuities will arise in future. Focusing on trends can produce a distorted picture, though, especially if trend selection is driven by conventional wisdom (or even military fads and fashions) rather than a thorough environmental scan. At worst, it can blind analysts to the potential for shocks: many predicted the persistence of the Soviet Union into the 1990s, for example, while analysis of the rise of China in our own time often assumes that past patterns of economic growth and military-political development will continue unbroken into the future.[9] But it is certainly possible to sketch out the key trends in today's conflict ecosystem, fit them into a broader analytical approach and thereby inform future planning in such a way as to improve our resilience to shocks and discontinuities.

The dragon and the snakes

In February 1993—only 14 months after the dissolution of the Soviet Union (one of Taleb's Black Swans)—James Woolsey, nominated by President Bill Clinton to be CIA director, told a congressional committee: 'We have slain a large dragon, but we live now in a jungle filled with a bewildering variety of poisonous snakes. And in many ways, the dragon was easier to keep track

---

8   Nassim Nicholas Taleb, *The Black Swan: The Impact of the Highly Improbable* (New York: Random House, Kindle Edition, 2011), Location 381ff.

9   See White, *How to Defend Australia* for an example of this.

of.'[10] He went on to describe a future operating environment—the post–Cold War era—characterised by threats from weak states, failing states and non-state actors such as terrorists, militias, narco-traffickers and pirates.

The collapse of communism was a classic discontinuity that caught future warfare analysts by surprise. But Woolsey's summation showed that at least some were able quickly to discern the event's longer-term impact—the way it altered the boundaries of the possible. Indeed, Woolsey's analysis proved to be an extremely prescient view of the conflict environment in the 1990s, which included a string of operations in the former Yugoslavia, Somalia, Rwanda, Cambodia, Sierra Leone and a host of other locations, all of them against primarily non-state actors or in response to post–Cold War chaos and state weakness.

In this context, the US military's dominance of conventional conflict against other nation-states, demonstrated so forcefully in the 1991 Gulf War, proved of borderline utility, even as its traditional great power adversaries—China and Russia—posed little threat in the immediate post–Cold War period. The US 'system-of-systems' approach, relying on high-tech, high-precision and extraordinarily expensive capabilities to defeat an enemy through direct, force-on-force engagement in a narrowly defined battlespace turned out to be ill-suited for the more diffuse and less combat-intensive peacekeeping, conflict prevention, reconstruction and humanitarian missions of the 1990s.

In 2001 another classic Black Swan—the September 11 Al Qaeda attacks—triggered the War on Terror, the invasions of Afghanistan and Iraq and a series of interventions in Africa, the Middle East, and South and South-East Asia. The US and its allies (including Australia) quickly became bogged down in counterterrorism, counterinsurgency and stabilisation missions, to such an extent that our conceptual and strategic bandwidth available to deal with other challenges—in particular, the rising military and economic strength of rival great powers including Russia and China, and the threats posed by Iranian and North Korean nuclear programs, cyberwarfare and state-sponsored asymmetric warfare—was severely limited.

---

10   Select Committee on Intelligence, *Hearing before the Select Committee on Intelligence of the United States Senate, One Hundred Third Congress First Session, Nomination of R. James Woolsey to Be Director of Central Intelligence* (Washington, DC: US Government Printing Office, 1993),76, www.intelligence.senate.gov/sites/default/files/hearings/103296.pdf.

At the same time, a series of workarounds dating back to the 1990s, in which capabilities originally designed for high-end conventional combat were twisted to fit the needs of low-intensity conflict, became increasingly mismatched to the missions that modern military forces faced. State-of-the-art aircraft, missiles, warships and conventional ground forces proved hugely expensive and largely ineffective against rag-tag irregular adversaries able to hide in complex human and physical (increasingly urban) terrain.

This mismatch became so obvious by 2005 that it prompted conceptual adaptations such as the counterinsurgency revival and the rediscovery of advisory and security force assistance missions for general-purpose forces.[11] It also drove organisational change such as the evolution of 'hyper-conventional' special operations forces (SOF) optimised for manhunting, raiding and integration with relatively cheap, low-tech air platforms including Predator and Reaper unmanned aerial vehicles (UAVs), AC-130 gunships and light attack aircraft like the A-29 Super Tucano or OV-10G Bronco.[12]

But all these were adaptations within the increasingly ill-fitting 'boundaries of the possible' established after the Cold War. Two further discontinuities exposed the weakness of this paradigm: the 2008 global financial crisis and the 2011 Arab Spring. The financial crisis underlined the war-weariness of Western publics unwilling to continue footing the bill for large-scale, long-duration, inconclusive wars of occupation in the Middle East, or to fund substantial foreign assistance and reconstruction amid an economic crisis at home. It also indirectly (through the inflationary effects of Western economic stimulus) exacerbated the social and economic stresses across the

---

11   Examples include the US Security Force Assistance Brigades, British Special Operations Brigade and the Australian Defence Force's Joint Task Force 629. See US Army, *Security Force Assistance Brigade: Operational and Organizational Concept* (2018), web.archive.org/web/20230714212641/https://fort benningausa.org/wp-content/uploads/2018/04/TCM_SFAB_2018.pdf; British Army, 'Special Operations Brigade', www.army.mod.uk/who-we-are/formations-divisions-brigades/6th-united-kingdom-division/army-special-operations-brigade/ [site discontinued]; Australian Department of Defence, 'JTF 629 Graduates First Class in the Philippines', *Defence News*, 14 November 2017, web.archive.org/web/202208 15091105/https://news.defence.gov.au/media/stories/jtf-629-graduates-first-class-philippines.

12   See Richard Sisk, 'A-29 Ground Attack Planes Tally More Than 260 Sorties in Afghanistan', *Military.com*, 5 May 2016, www.military.com/daily-news/2016/05/10/a29-ground-attack-planes-tally-more-260-sorties-afghanistan.html; Tyler Rogoway, 'Those Old OV-10 Broncos Sent to Fight ISIS Were Laser Rocket-Slinging Manhunters', *The War Zone*, 17 May 2016, www.thedrive.com/the-war-zone/3519/those-old-ov-10-broncos-sent-to-fight-isis-were-laser-rocket-slinging-manhunters.

Middle East and North Africa that eventually triggered the Arab Spring.[13] Western governments reacted by seeking a 'light footprint' model of war that repurposed earlier adaptations developed for Iraq and Afghanistan— SOF, airstrikes, drones, mentoring of local government allies and non-state proxies—as a way of 'leading from behind' to enable disengagement from the region.

The intervention in Libya in 2011, the spillover of conflict from Libya into the rest of northern and western Africa, the rise of Islamic State, the war in Yemen and the re-engagement of coalition forces in Iraq and then Syria showed how difficult disengagement would be. It also underlined the return of state adversaries—peer and near-peer competitors—to the operating environment. In 2013 Russia and Iran seized the initiative in Syria, in 2014 Russia annexed Crimea and invaded Ukraine, and in 2015 Russia and Iran deployed full-scale combat units to Syria even as Tehran and Washington signed a nuclear deal that gave Iranian forces increasing freedom of action across the region.

The 'post-Woolseyan' environment

After two decades of post–Cold War snakes, so ably described by Jim Woolsey, it is now clear that the dragon is back—we are in a 'post-Woolseyan' security environment. In this environment, Western powers are dealing with both state and non-state threats, at the same time and in many of the same places. State actors (principally Russia, Communist China, Iran and North Korea) have learned from non-state threats over the past 20 years, adopting modular organisations, multirole platforms, swarm tactics and improvised offensive and defensive weapon systems that mimic many of the techniques evolved by non-state adversaries as ways to confront conventionally superior coalition forces in Iraq and Afghanistan. They have taken advantage of Western countries' tunnel vision on terrorism (and focus on extricating ourselves from a Middle Eastern morass of our own making) to enhance nonconventional and asymmetric capabilities, improve their economic and geopolitical positioning, and re-engage as great powers

---

13   See Adeel Malik and Bassem Awadalla, *The Economics of the Arab Spring*, CSAE Working Paper WPS/2011-23 (Oxford: Centre for the Study of African Economies, 2011); Michael Gordon, 'Forecasting Instability: The Case of the Arab Spring and the Limitations of Socioeconomic Data', *Insight and Analysis*, 8 February 2018, www.wilsoncenter.org/article/forecasting-instability-the-case-the-arab-spring-and-the-limitations-socioeconomic-data.

with global reach. Conventional capabilities—Chinese anti-ship ballistic missiles, submarines and aircraft carriers, advanced Russian armoured vehicles, hypersonic missiles—have been combined with improvements to nuclear arsenals and investment in nonconventional capabilities such as cyberwarfare, bioweapons, human performance enhancement, space warfare systems and improved capacity for economic warfare.

At the same time, the explosion in electronic connectivity noted earlier— along with a proliferation of consumer handheld smart devices, global positioning system (GPS)–enabled systems, additive manufacturing, and the fact that most conflict now occurs in and around urbanised areas and involves populations with a high degree of technological and mechanical skill—has allowed non-state armed groups to achieve levels of lethality and precision that were once the sole preserve of nation-states. Precision-indirect fire systems controlled through Google Earth using smartphones and networks of linked observers sharing targeting data via mobile phone, teleoperated sniper weapons, grenade-carrying 'drone bombers' employing repurposed hobby drones, 3D-printed firearms, precisely machined explosively formed penetrators (EFPs) and a host of other advanced capabilities are now available to any non-state armed group with the inclination and technical skill to exploit them. Groups can exploit electronic connectivity to actively share knowledge or passively copy and learn from each other. Some, notably Islamic State, have used conventional Western-style tactics (and acquired tanks, mobile artillery and several working helicopters) only to relearn the lessons of 1991—that confronting US forces (or US allies) in the open, conventionally arrayed, is a recipe for disaster. Others have focused on amorphous modular or cell-based networks that are optimised for blending into complex human and physical terrain.

In both cases, these can be seen as evolutionary responses to Western dominance of one particular narrowly defined form of 'conventional' combat since the Cold War. Different threat actors—state and non-state— are adapting to that dominance by seeking remarkably similar ways to avoid our conventional strength, developing techniques and technologies that offset our advantages, and attempting to invalidate our high-tech, battlefield-centric way of warfare. We find ourselves in a crowded, cluttered, highly connected, predominantly urban and coastal environment, operating against a mix of state and non-state actors, all of whom are applying irregular methods designed to overwhelm us with a massive number of small challenges rather than a single overwhelming threat.

Boundaries of the possible

This suggests certain boundaries of the possibility for future warfare. These are neither predictions nor projections from current conditions. Rather, they are observations of some of what seem to be the more enduring elements of the current conflict environment. Thus, they represent characteristics we might expect to see in the threat environment of the next decade or so, irrespective of which adversary we find ourselves fighting. They can be summarised as follows:

Nonlinearity. Actors in the future conflict environment are likely to avoid linear approaches (such as the deep/close/rear spatial model of the battlespace, or the chronologically sequential phasing construct in operational planning).[14] Instead, they will seek to influence all elements of an adversary's operational system, across the full breadth and depth of a spatial and temporal matrix, targeting identified vulnerabilities in key political, military, economic, social, infrastructural and informational nodes. They will flexibly adapt their approach on a continuous basis throughout a conflict, altering the means used (kinetic/non-kinetic, lethal/nonlethal, cyber/physical) while jumping among multiple domains (air, space, sea, land, the electromagnetic spectrum, information and cyberspace) to avoid countermeasures, and adaptively shifting focus to keep adversaries off-balance and prevent them concentrating force against any one target.

Simultaneity. Actors in the conflict environment may seek to generate simultaneous effects across an adversary's entire civil–military system (as described above) in order to create a bandwidth challenge that saturates the adversary's ability to focus on any one challenge. The goal will be to overwhelm an adversary cognitively (by preventing an enemy from detecting, isolating and responding to any one threat), physically (by soaking up assets that could otherwise be concentrated on a single attack) and politically (by creating distraction and discord that limits an adversary's capacity for coherent action).

Liminality. Conflict actors are likely to recognise that pervasive electronic communications and intelligence, surveillance and reconnaissance capabilities now render truly clandestine or covert operations increasingly

---

14 For a critique of the latter see Lauren Fish, 'Painting by Numbers: A History of the U.S. Military's Phasing Construct', *War on the Rocks*, 1 November 2016, warontherocks.com/2016/11/painting-by-numbers-a-history-of-the-u-s-militarys-phasing-construct/.

unachievable. Instead, they will seek ambiguity—obfuscating the nature of their activities and obscuring the identity or status of key combat elements to make it harder for traditional (conventional) conflict actors to respond. They will manage their detectable signature (across all domains) to remain below an adversary's detection threshold until the last possible moment, then pop up above that threshold for limited periods of time only, to achieve specific tactical and operational goals, before reducing their signature to slip below the threshold before an adversary can marshal a response.

Decisive shaping. Recognising the overwhelming superiority of US and allied forces within the narrow confines of a conventional combat engagement, adversaries will seek to achieve campaign objectives during the pre-combat ('shaping') stage of a campaign. They will use economic, political, cyber and unconventional warfare techniques to ensure mission success before the first airstrike or bombardment is launched, or the first combat unit crosses the line of departure, making the shaping stage the decisive element in a campaign rather than (as is traditional) the combat phase. As a consequence, where an actor is successful in achieving goals through shaping from below the threshold of open conflict, some campaigns may end in a decisive victory for that actor, while never progressing to an overt combat phase at all. In this form of campaign, conventional actors may thus already be defeated by the time they realise they are in a conflict.

Conceptual envelopment. Adversaries may expand (and some, notably China, have already expanded) their concept of war far beyond the narrow definition used by Western war colleges. 'Non-military and trans-military war operations' as defined by Chinese thinkers in the 1990s, and embodied in China's 'Three Warfares Doctrine' and more recent analyses by Chinese strategists like Zhang Shibo, may include activities such as control of scarce commodities or manufacturing capabilities, strategic real estate acquisition, technology theft, commercial and financial warfare, drug warfare, cyberwarfare, 'public opinion warfare', manipulation of legal norms ('lawfare'), currency manipulation and biowarfare.[15] One resulting risk is that of conceptual envelopment—where Western decision-makers engage in actions they consider normal peacetime interaction but which

---

15 See Qiao Liang and Wang Xangsui, *Unrestricted Warfare* (Beijing: PLA Literature and Arts Publishing House, 1999); Elsa Kania, 'Weaponizing Biotech: How China's Military Is Preparing for a "New Domain of Warfare"', *Defense One*, 14 August 2019, www.defenseone.com/ideas/2019/08/chinas-military-pursuing-biotech/159167/; Zhang Shibo, *New Highland of War* (Beijing: Chinese National Defense University Press, 2017).

an adversary with a broader concept considers acts of war; alternatively, an adversary may engage in trans-military or non-military war against us, while we remain unaware of that fact until too late.

Democratisation of lethality. Finally, the future warfare environment is likely to include a wider range of actors with access to a more capable array of high-end lethal technologies. These may be 'onboard' capabilities embedded within combat elements or carried at unit or formation level within a force, or distributed capabilities that combat actors can access through collaborative engagement techniques, dispersed operations or access to artificial intelligence capabilities, allowing them to reach back for remotely held support. In the case of state actors, this is likely to translate into increased levels of precision, lethality and manoeuvrability down to the small-team level (sections, combat pairs or even enhanced individuals). In the case of non-state actors, it will be manifested in individuals and small teams deploying levels of lethality that were previously restricted to large groups or government organisations. Space-based and satellite capabilities, artificial intelligence, additive on-site manufacturing, 3D printing of energetics (such as explosives and incendiary materials) and the proliferation of autonomous systems—including UAVs and counter-drone systems— will give land forces the ability to operate simultaneously in multiple domains at a far higher level of lethality but in smaller and more dispersed combat groups.

Conclusion: Black Swans in future war

These characteristics, as noted, represent an attempt to make explicit the likely outside parameters—boundaries of the possible—for the future warfare environment as we currently understand it. But as noted earlier, the role of shocks and discontinuities, Taleb's Black Swans, must not be discounted. Indeed, Taleb's list of shocks quoted above (and the additional instances mentioned such as pandemics, revolutions and national disasters) make it clear that such events account for the most significant changes in the character of war, and in our understanding of the conflict environment. All of this suggests a key conclusion relevant to our attempts to understand future war, namely that *predicting* the nature of the future conflict environment may not be the appropriate goal. Even more so, using a selective assessment of current conditions, combined with an incomplete understanding of the key inputs that shape those conditions, can lead to straight-line projections that dramatically misperceive the likely future environment, actively harming our ability to make sense of it.

Instead, a more appropriate approach may be to map the boundaries of the possible future warfare but then recognise the inevitability of shocks, discontinuities and strategic surprises that shift those boundaries in unpredictable ways. As a consequence, any assessment of this kind can be nothing more than a hypothesis. Having developed such a hypothesis (and elaborated it to the degree possible using current data) the proper role of a future warfare analyst should then be to look for data in the environment that tend to disprove, modify or invalidate that hypothesis—in the manner of an indicators and warnings (I&W) problem in strategic intelligence assessment.

This approach would treat the future of warfare as a natural live experiment, using an initial hypothesis to develop a set of observable indicators to help analysts identify when and how the environment is diverging from that hypothesis. Analysts would then track these indicators and continually update the 'boundaries of the possible' based on I&W observations. The goal of such an approach would not be to predict the specific details of future wars, but rather to frame the range of possibilities within which any future conflict would be most likely to occur. This understanding would inform capability acquisition and planning decisions, so that when discontinuities do inevitably occur, the organisation's capabilities represent the closest possible match to reality, while plans and concepts are sufficiently developed (and updated) to enable an agile response to surprise.

Nothing in this approach can—or should be expected to—avoid the surprises and shocks that result from Black Swans. There will always be discontinuities that drive changes in our understanding of the character of war, and some of these—the Russian Revolution, Pearl Harbor, Hiroshima, the 9/11 attacks, the Arab Spring—will catch us by surprise and require massive and rapid reorientations. But a more modest approach to prediction, one that seeks to understand the range of possibilities for, and characteristics of, future conflict, rather than attempting to project its details from current conditions, may ironically make us more resilient in the face of such shocks.

References

Australian Department of Defence. 'JTF 629 Graduates First Class in the Philippines'. *Defence News*, 14 November 2017. web.archive.org/web/20220815091105/https://news.defence.gov.au/media/stories/jtf-629-graduates-first-class-philippines.

British Army. 'Special Operations Brigade'. www.army.mod.uk/who-we-are/formations-divisions-brigades/6th-united-kingdom-division/army-special-operations-brigade [site discontinued].

Chiefs of Staff Committee. 'An Appreciation of the Strategical Position of Australia'. In *A History of Australian Strategic Policy since 1945*, edited by Stephan Frühling, 53–98. Canberra: Department of Defence, 2009.

Dibb, Paul. *Review of Australia's Defence Capabilities*. Canberra: Australian Government Publishing Service, 1986. www.aspistrategist.org.au/wp-content/uploads/2022/02/Review-of-Australias-Defence-Capabilities-1986.pdf.

Ducote, Brian M. 'Challenging the Application of PMESII-PT in a Complex Environment'. Student monograph, School of Advanced Military Studies, Fort Leavenworth, Kansas, 2010. apps.dtic.mil/sti/pdfs/ADA523040.pdf.

Fish, Lauren. 'Painting by Numbers: A History of the U.S. Military's Phasing Construct'. *War on the Rocks*, 1 November 2016. warontherocks.com/2016/11/painting-by-numbers-a-history-of-the-u-s-militarys-phasing-construct/.

Gordon, Michael. 'Forecasting Instability: The Case of the Arab Spring and the Limitations of Socioeconomic Data'. *Insight and Analysis*, 8 February 2018. www.wilsoncenter.org/article/forecasting-instability-the-case-the-arab-spring-and-the-limitations-socioeconomic-data.

Head Modernisation and Strategic Planning. *Adaptive Campaigning 09: Army's Future Land Operating Concept*. Canberra: Directorate of Army Research and Analysis, 2009. researchcentre.army.gov.au/sites/default/files/acfloc_2012_main.pdf.

Kania, Elsa. 'Weaponizing Biotech: How China's Military Is Preparing for a "New Domain of Warfare"'. *Defense One*, 14 August 2019. www.defenseone.com/ideas/2019/08/chinas-military-pursuing-biotech/159167/.

Kilcullen, David. *Complex Warfighting*. Canberra: Australian Army, 2005. indianstrategicknowledgeonline.com/web/complex_warfighting.pdf.

Kilcullen, David. *Out of the Mountains: The Coming Age of the Urban Guerrilla*. New York: Oxford University Press, 2013.

Kuhn, Thomas S. *The Structure of Scientific Revolutions*, 50th anniversary edition. Originally published 1962, reprinted Chicago: The University of Chicago Press, 2012.

Malik, Adeel, and Bassem Awadalla. *The Economics of the Arab Spring*. CSAE Working Paper WPS/2011-23. Oxford: Centre for the Study of African Economies, 2011.

Porter, Patrick. 'The World Is Not Flat: War and Distance in the Twenty First Century'. In *Geo-strategy and War: Enduring Lessons for the Australian Army: The 2015 Chief of Army History Conference*, edited by Peter Dennis, 313–324. Canberra: Big Sky Publishing, 2015. web.archive.org/web/20200403225438/https://www.army. gov.au/sites/default/files/2015_geo_strategy_and_war.pdf?acsf_files_redirect.

Qiao Liang and Wang Xangsui. *Unrestricted Warfare*. Beijing: PLA Literature and Arts Publishing House, 1999.

Rogoway, Tyler. 'Those Old OV-10 Broncos Sent to Fight ISIS Were Laser Rocket-Slinging Manhunters'. *The War Zone*, 17 May 2016. www.thedrive.com/the-war-zone/3519/those-old-ov-10-broncos-sent-to-fight-isis-were-laser-rocket-slinging-manhunters.

Select Committee on Intelligence. *Hearing before the Select Committee on Intelligence of the United States Senate, One Hundred Third Congress First Session, Nomination of R. James Woolsey to Be Director of Central Intelligence*. Washington, DC: US Government Printing Office, 1993. www.intelligence.senate.gov/sites/default/ files/hearings/103296.pdf.

Sisk, Richard. 'A-29 Ground Attack Planes Tally More Than 260 Sorties in Afghanistan'. *Military.com*, 5 May 2016. www.military.com/daily-news/2016/ 05/10/a29-ground-attack-planes-tally-more-260-sorties-afghanistan.html.

Stevenson, Robert. 'The Tyranny of Distance: Geo-strategy and the New Guinea Campaign of 1914'. In *Geo-strategy and War: Enduring Lessons for the Australian Army: The 2015 Chief of Army History Conference*, edited by Peter Dennis, 19–56. Canberra: Big Sky Publishing, 2015. web.archive.org/web/20200403225438/ https://www.army.gov.au/sites/default/files/2015_geo_strategy_and_war.pdf? acsf_files_redirect.

Taleb, Nassim Nicholas. *The Black Swan: The Impact of the Highly Improbable*. New York: Random House, 2011.

US Army. *Security Force Assistance Brigade: Operational and Organizational Concept*. 2018. web.archive.org/web/20230714212641/https://fortbenningausa.org/wp-content/uploads/2018/04/TCM_SFAB_2018.pdf. *How to Defend Australia*. Melbourne: Latrobe University Press, 2019.

Zhang Shibo. *New Highland of War*. Beijing: Chinese National Defense University Press, 2017.

4

Space: Ambiguity, Vulnerability and a Changing Character

Mark Hilborne

Human activity is becoming increasingly reliant on space technology.[1] This extends from the transmission of television and radio signals to providing time synchronisation, which is central to the functioning of critical national infrastructure. In parallel, space technology provides militaries with a powerful set of capabilities, including intelligence gathering, precision navigation and targeting and communications. As a result of the advantages and dependencies that are becoming increasingly entrenched, space is developing into a domain that is increasingly contested, and a domain that states will seek military advantage from and within. As such, contemporary military strategy needs to carefully consider the security advantages and threats presented and how to gain or maintain advantage.

While neither space nor its militarisation is a new area of human activity, the increase in the capabilities derived from space, and the attendant disaster that would accompany their loss, identifies space security as a particularly urgent issue. There is an increasing number of both state and commercial actors in space, placing more and more assets there. This drives pressure on the governance of space, as orbital slots and the allocation of radio

---

1   Editors' note: as mentioned in the Introduction, this chapter was written in 2020, and global events since then may have overtaken some aspects. As the pace of technological and cultural change has continued to accelerate, the editors have opted to present this text as drafted to minimise further delays in publication.

frequencies become congested. More worrying though, given the critical nature of space assets, is an increasing likelihood of these being targeted by an adversary, and this raises the spectre of conflict in space.

Looking forward to future conflict scenarios involving space, there is clearly a military aspect, but also a non-military element, and the latter will impact the former. Developments in commercial space are driving innovation and numbers. This will affect military uses of space in that much technology will increasingly be developed by commercial companies. Military forces will need to look to and leverage these commercial companies for solutions. These commercial actors too will change the dynamics of space. For instance, if information is power, and commercial companies become the dominant force in data collection, they will be increasingly important in equations of space power. Furthermore, this will impact operational security for military forces—for as commercial satellite imagery (CSI) expands, there will simply be nowhere to hide.

These developments will go hand-in-hand with the more traditional state-centric models of competition and conflict. A number of states are overtly developing counterspace capabilities. These programs will intensify existing sensitivities in space, particularly on behalf of the US, whose investment in and dependence on space is far beyond any other nation. This in itself produces a wariness of competitors. However, the existence of what is considered a vulnerability gap may aggravate this, provoking some opponents to carry out a pre-emptive strike on US space assets to generate an initial asymmetrical advantage in a conflict. Overlaid with America's dual-use assets for nuclear and non-nuclear commands, this may easily lead to misinterpretation of intent and thus escalation. The combination of these dynamics creates an unstable fabric, exacerbated by the difficulty in attribution in space and the hazy regulations of what is permissible in space. This yields a natural environment for subthreshold activities. In the absence of any significant efforts to allay these trends, the prognosis for avoiding conflict in space is not optimistic.

The developing context

During the Cold War, the key antagonists were balanced via a bilateral distribution of power. This was true in both space technology and more widely, with both the US and the USSR possessing broadly similar capabilities and numbers. During the Cold War, the space capabilities they wielded were seen in many respects as stabilising, as they could be used to monitor the arms control treaties and allow verification. This symmetry changed abruptly with the dissolution of the Soviet Union, whose development in space began to atrophy.

Accompanying this was a change to the composition of space activity. Prior to 1990, over 90 per cent of all satellites launched into space came from the US or USSR, almost three-quarters of which served a classified military purpose. This began to change dramatically after the end of the Cold War, with a sharp increase in commercial ventures in space. While some hail Gulf War I as the beginning of a new space age, in fact, a better categorisation of that conflict is that it was the culmination of the first space age, with the second typified by the rise of the commercial sector.

These changes affect the dynamics of space, from one that was underpinned by arms balancing, to an environment that is far more complicated and pluralistic, in which the measurement of space power is more difficult and in which the technological edge held by the most advanced militaries are eroded by commercial actors in a number of ways.

Commercial sector and NewSpace

The rise of the commercial sector is already having a significant effect on the complexion of space activities, and this trend looks likely to increase. While commercial space is not a new phenomenon, it has become responsible for the bulk of space activities currently. Adding to the changing dynamics is the entry of so-called NewSpace into the market.[2] These developments, while bringing many advantages to civilian applications, will also have strategic ramifications. These will relate to technological innovation and advantage, and to military operational security. Most innovation today in space, and indeed in other areas, takes place in the commercial sector, not government labs, in stark contrast to previous generations. US and other Western militaries will need to pay close attention to these developments, and where possible, absorb and exploit the advances and advantages that the commercial sector develops.

---

2   NewSpace is often considered to be a more entrepreneurial element of commercial space. These companies are self-funded, and do not rely on governments for support, and often work on much smaller budgets than more established space companies.

An examination of the number of active satellites and satellite launches, as well as their value, provides some indication of the impact of this rising sector. The Space Report indicates that the commercial sector is now approximately 80 per cent of the space economy by value.[3] As of November 2019, of the 901 US satellites in orbit, 523, or 58 per cent, are commercial satellites.[4] Dwarfing these numbers, new commercial constellations are being launched. The most prolific of these is SpaceX, with its Starlink constellation. Sixty Starlink satellites have so far been launched, while there are plans to launch hundreds, or perhaps over a thousand more in 2020. SpaceX has had an application for 12,000 more launches approved and has submitted applications for an additional 30,000. This contrasts with the number of active satellites in orbit prior to the Starlink launches of approximately 1,900.[5]

With a view to future warfare, while the mere fact that numbers are increasing to the point of overshadowing military satellite numbers is significant, it is the strategic implications that are of particular interest. One of the most profound areas where this will have an effect is commercial satellite imagery (CSI). While the capability and number of active CSI satellites continue to escalate, the model of CSI is well established. An early example of such an operation is the French *Satellite pour l'Observation de la Terre*, or SPOT, first launched in 1986 by the *Centre national d'études spatiales* (CNES). This operation allows affordable and near real-time imagery of good resolution. The initial resolution of 10 m for SPOT 1 has been gradually enhanced such that the latest SPOT 7 offers a 1.5 m resolution. The cost for an image can be in the region of US$1,000–3,000. Such imagery can allow potential adversaries—state and non-state actors alike—without their own dedicated space capabilities to collect information about US or other Western military forces' positions and movements along with the deployment of specific equipment. This can be juxtaposed with other freely available sources, such as Google Maps and Street View, to provide further information. Allied with the broader transparency of Western states, political and military objectives may be gauged from media commentaries and press briefings.

3   Space Foundation, 'Commercial Space Revenue Climbs to All-Time High of $328.86 Billion, the Space Report Reveals in Q3 Analysis', *Space Foundation*, 31 October 2019, www.spacefoundation. org/2019/10/31/commercial-space-revenue-climbs-to-all-time-high-of-328-86-billion-the-space-report-reveals in q3 analysis/.
4   Union of Concerned Scientists, *UCS Satellite Database*, 8 Dec 2005, www.ucsusa.org/resources/satellite-database.
5   Union of Concerned Scientists, *UCS Satellite Database*.

The implications of CSI on military operations have attracted attention in the past. The US Army's *Field Manual 100-5, Operations, 1986* notes that advances in strategic surveillance technology make it:

> increasingly more difficult for the U.S. to mask or cloak any large-scale marshalling or movement of personnel and equipment. Since transparency may seriously jeopardize the military's ability to achieve strategic or tactical surprise, or worse make surprise highly unlikely, the U.S. military may have to change its methods for achieving surprise.[6]

In the late 1980s there was a study undertaken by the Carnegie Endowment for International Peace, assessing the military utility of Landsat, SPOT and Soyuzkarta KFA-1000 imagery.[7] Even though the 10-metre resolution of the early SPOT was not considered the most advanced, and would soon be outclassed, it was nonetheless sufficient for the study group to enable the targeting-related tasks contained in their objectives. The Carnegie study concluded that CSI is 'rich in information which can be used to affect the planning and execution of military operations'.[8] Adversaries could use CSI for a number of objectives, from targeting, as noted in the Carnegie study, to influencing decision-makers, to affecting their opponent's military operations and tempo. A later Carnegie study noted a more profound impact—a change in the suppliers of information that:

> shifts power from the former holders of secrets to the newly informed, [which] has implications for national sovereignty, for the ability of corporations to keep proprietary information secret, and for the balance of power between state and nonstate actors.[9]

---

6   US Army Headquarters, *Field Manual 100-5, Operations, May 1986* (Washington, DC: Department of the Army, 1986), 176, Appendix A.

7   See Peter D Zimmerman, 'Introduction to Photo-Interpretation of Commercial Observation-Satellite Imagery', in *Commercial Observation Satellites and International Security*, ed. Michael Krepon (Basingstoke: Macmillan in association with the Carnegie Endowment for International Peace, 1990). The Soviet Union had entered into the CSI market in 1987, through a trading company V/O Soyuzkarta, offering imagery of 5 m resolution from the Soyuzkarta KFA-1000. Its resolution was some of the highest then available, though there were some shortcomings and inexpert marketing. See also Directorate of Intelligence, *Soviet Commercial Space Photography: Offering Resolution as the Solution* (Washington, DC: Directorate of Intelligence, 1990), declassified 1999, nsarchive2.gwu.edu/NSAEBB/NSAEBB501/docs/EBB-47.pdf. Also regarding Landsat and SPOT, see Micheal Krepon, 'Peacemakers or Rent-a-Spies?' *Bulletin of the Atomic Scientists* 45, no. 7 (September 1989): 12–15, doi.org/10.1080/00963402.1989.11459716.

8   Zimmerman, 'Introduction to Photo-Interpretation of Commercial Observation-Satellite Imagery', 203.

9   Ann M Florini, and Yahya A Dehqanzada, *No More Secrets? Policy Implications of Commercial Remote Sensing Satellites* (Carnegie Endowment for International Peace, 1 July 1999), carnegieendowment.org/1999/07/01/no-more-secrets-policy-implications-of-commercial-remote-sensing-satellites-pub-150.

The ability to determine when satellites' orbits will pass overhead has provided a useful opportunity to avoid detection and allow clandestine activities, permitting significant actions to go undetected. For instance, in 1998, India was able to hide a number of telltale signs of an impending nuclear test from satellites, based on their knowledge of US satellite orbits, with the result that the US was taken by surprise by the test.[10] Over 20 years later, such opportunities are dwindling—the proliferation of CSI satellites and their increasing capabilities will reduce these options significantly, or possibly entirely. In June 2014, the US Department of Commerce responded to an application from DigitalGlobe, and lifted restrictions on more detailed satellite images, allowing an increase in resolution from 50 cm down to 31 cm. Such resolution allows features as small as 'mailboxes and manholes' to be identified, according to a DigitalGlobe spokesperson.[11] DigitalGlobe had requested the same reduction 15 years before and was refused. A Senate Select Committee on Intelligence that reviewed the later request noted that 'foreign commercial imagery providers may soon be able to provide imagery at or better than the currently allowed commercial U.S. resolution limit of 0.5 meters'.[12] The statement underlines the international competition in CSI and the limit of a state's jurisdiction in space. Restrictions on US companies hampered their potential while allowing others to flourish and may in fact provide an incentive for others to develop their capabilities.

In the months leading up to the decision to allow the higher-resolution imagery, National Reconnaissance Office (NRO) Director Betty Sapp stated her support for DigitalGlobe's request:

> The NRO has always been a very strong supporter of the commercial imagery guys. We share hardware, software, test equipment and new technology. We want to continue that in a much more fundamental way in the future … . We want to make sure whatever they do, we can take full advantage of. We want to make sure we're not doing anything they can do.[13]

---

10 It is suspected that when the US ambassador to India shared satellite imagery with Delhi in 1995 in order to dissuade them from testing, the Indians took steps to ensure they could operate unobserved after this. See James Risen, Steven Lee Myers and Tim Weiner, 'U.S. May Have Helped India Hide Its Nuclear Activity', *New York Times*, 25 May 1998, www.nytimes.com/1998/05/25/world/us-may-have-helped-india-hide-its-nuclear-activity.html.

11 'US Lifts Restrictions on More Detailed Satellite Images', *BBC News*, 16 June 2014, www.bbc.co.uk/news/technology-27868703.

12 Mike Gruss, 'U.S. Intelligence Community Endorses Company's Bid to Sell Sharper Imagery', *SpaceNews*, 18 April 2014, spacenews.com/40263us-intelligence-community-endorses-companys-bid-to-sell-sharper-imagery/.

13 Gruss, 'U.S. Intelligence Community Endorses Company's Bid'.

Such a statement suggests that the commercial companies have developed capabilities that have real utility to the intelligence community and that the latter will seek to leverage those where possible.

While the imagery itself is of tremendous value, making sense of the imagery, and in particular deriving useful data from it, is equally valuable. This may have been the driving force behind the decision by Google to purchase the start-up satellite imaging company, Skybox Imaging, in 2014, just as the US Government decided to permit the higher resolution.[14] Initial reports and Google's own early press statements suggested that the acquisition was driven by the desire to keep Google Maps up to date in near real-time and provide internet service.[15] However, the data-processing abilities of Skybox may have been the real intent behind the acquisition.

A comparison of Skybox's capabilities with traditional Earth observation satellites is illuminating. The latter tend to be large, expensive and use advanced techniques and technology. These factors limit the number in orbit. In contrast, Skybox achieved a platform that is cheap and light but utilises innovative image-processing software to countervail the hardware's shortcomings.[16] The company was able to launch after raising just US$91 million in total, demonstrating how accessible the technology has become.

The SkySat design also permits comparatively high data output. Currently, weather satellites might generate 60 gigabytes of raw data which, when processed become almost 6 terabytes of climate information, such as forecasts, cloud coverage, sea ice concentrations and surface temperatures.[17] By comparison, Skybox was able to mine 1 terabyte of preprocessed data daily in their early phase of operation with only a few satellites. Once processed, the potential is vast. Given the affordable nature of the satellites, and the resources of a company such as Google, one observer hypothesised

---

14   Skybox was renamed Terra Bella on 8 March 2016. They were also acquired by Planet Labs from Google in 2017.

15   Jessica Guynn, 'Google Buying Skybox Satellite Company to Boost Maps', *USA Today*, 10 June 2014, eu.usatoday.com/story/tech/2014/06/10/google-buying-skybox-satellite-company/10284045/.

16   The SkySat-1 satellites are 83 kg and are approximately 600 mm × 600 mm × 800 mm in dimension. 'SkySat Constellation', Satellite Mission Catalogue, *eoPortal*, directory.eoportal.org/web/eoportal/satellite-missions/s/skysat.

17   Alex Knapp, 'Forecasting the Weather with Big Data and the Fourth Dimension', *Forbes*, 13 June 2013, www.forbes.com/sites/alexknapp/2013/06/13/forecasting-the-weather-with-big-data-and-the-fourth-dimension/. Knapp is discussing the Joint Polar Satellite System (JPSS) Common Ground System.

that the result could be continuous real-time coverage of the entire planet, creating 'data sets that seem like science fiction—data sets of the kind that were formerly available only to the NSA, and then only in theory'.[18]

While the commercial value of these capabilities is untold, advances in commercial satellite technology have strategic ramifications and will impact future warfare. Who has the lead in space—who has information supremacy? The technological advantage held by the military as the leader of information flowing from space is likely to be eroded, and in the near future, it may be eclipsed. Both state actors and non-state actors could access CSI imagery to collect intelligence, conduct industrial espionage and plan military or terrorist action.[19] On the other hand, the mass availability of CSI may enable a number of functions that would be costly and difficult from a purely military perspective. Additionally, CSI, as well as commercial space situational awareness capabilities, can contribute to growing global transparency. States will need to accept an era where there is universal observation, exploiting the positive impacts, while also mitigating the negative.

State competition

None of these dynamics mean that state-on-state competition is slowing. Among the challenges in space, this aspect of competition is still a significant dynamic. Reliance on space is so critical that space-based assets are now potentially a prime target. The inevitable result is competition and potential friction in space between spacefaring nations, alongside the development of counterspace technologies. It is abundantly clear that the development of advanced space-derived capabilities by the US has generated significant advantages in the military domain. These extend to targeting, communication and intelligence, as well as the operation of remotely piloted air systems. Chinese and Russian military leaders fully comprehend

---

18   See Will Oremus, 'Google's Eyes in the Sky', *Slate*, 13 June 2014, slate.com/technology/2014/06/google-skybox-titan-aerospace-acquisitions-why-it-needs-satellites-and-drones.html. Also see David Samuels, 'Inside a Startup's Plan to Turn a Swarm of DIY Satellites into an All-Seeing Eye', *Wired*, 18 June 2013, www.wired.com/2013/06/startup-skybox/.

19   Notably, during Gulf War I, both sides accessed commercial imagery. The US utilised both Landsat and SPOT imagery, even though it was of lower resolution than military satellites, to aid analysis and planning, while Iraq accessed SPOT satellite imagery prior to its invasion of Kuwait. France quickly cut off access after the invasion, however. Cynthia AS McKinley, *When the Enemy Has Our Eyes* (Columbus: Biblioscholar, 2012), 21.

the unique information advantages afforded to the US by space systems and are developing capabilities to deny the US use of space in the event of a conflict.

These are trends that could have a profound impact on the ways in which wars are fought, and perhaps why they are fought. They will also have an impact on the balance of power and on the ability to exploit space for the host of functions that it currently affords, and potentially many more that have not yet been conceived. Any kind of conflict in space could have catastrophic results for the long-term use of space.

While the US has a distinct advantage in space, other spacefaring nations are making notable advances, many of which are counterspace technologies. India, for instance, tested its Reusable Launch Vehicle Technology Demonstrator (RLV-TD). This may potentially permit the development of a reusable lifting body, and some argue it could also be the basis of boost-glide hypersonic weapons. In February 2017, India launched 107 satellites from a single rocket, a record number. It also tested an anti-satellite weapon (ASAT) in April 2019. Russia too is fielding a number of advanced capabilities in space, including restarting some of its Cold War counterspace programs. Most notable is its Nudol A-235 direct-ascent ASAT, which has so far undergone four separate tests and one actual interception. There have also been a series of rendezvous and proximity operations since 2013, some close to Western satellites in geostationary orbit.[20]

But most attention in recent years has been on China. Its achievements in space have been spectacular, including a manned space station and meaningful plans for a joint lunar base with Russia, and such achievements continue unabated. China demonstrated its counterspace potential with its 2007 ASAT test, which created a massive debris field, thus simultaneously demonstrating the dangers of such weapons. Its broader manned spaceflight program has proven its proficiency, having built a space station and conducted docking manoeuvres. In June 2016, China launched its most powerful rocket, the Long March 7. Though the launch was a notable achievement in itself, the secondary payload it carried—the Aolong-1 or 'Roaming Dragon', a small satellite designed to collect space debris with a

---

20   Brian Weeden and Victoria Samson, eds, *Global Counterspace Capabilities: An Open Source Assessment* (Washington, DC: Secure World Foundation, April 2018), 2-2, swfound.org/media/206118/swf_global_counterspace_april2018.pdf. Also Mike Gruss, 'Russian Satellite Maneuvers, Silence Worry Intelsat', *SpaceNews*, 9 October 2015, spacenews.com/russian-satellite-maneuvers-silence-worry-intelsat/.

robotic arm—has provoked familiar speculation about the true nature of China's space program. While ostensibly a useful technology, such a device could clearly be used in a malicious manner.

Similarly unclear was a test in 2013, in which China launched a rocket on an apparent high-altitude test mission. While of course plausible, a number of analysts suggest that China appeared instead to be testing a kinetic interceptor launched by a new rocket that could reach geostationary orbit. If true, this is a notable development, as no other country has yet developed such a capability.[21]

These latter examples identify a key theme in space operations. Many, if not most, space technologies are dual use or have dual-use potential. This creates an inherent uncertainty of intent. Military secrecy limits transparency and the prospect of misperceptions and mistrust relating to military activities in space increases. The reduction in stability raises the possibility of conflict in space, or of space being a trigger for tensions below.

Where the potential for misperceptions and mistrust is most critical—though not exclusively—is the relationship between the US and China. The US is highly sensitive to developments in space due to its investment in and reliance on space, as well as the symbolic value of space in the bilateral competition with the USSR during the Cold War. With space so central to America's strategic mindset, China's growth in space, allied to its opaque politics, is bound to increase Washington's concern.

US interests and sensitivities

In order to manage stability in space, the interests of the US cannot be overlooked. That is not to say that the US has a particular prerogative in space, but rising powers will need to be conscious that their actions will create reactions. This is partly due to the place that space has in both US strategic and popular culture, resulting from its triumphant space endeavours during the Cold War, which serve as an important symbol of America's scientific prowess and global leadership. It is also the result of the

---

21   Brian Weeden, 'Through a Glass, Darkly: Chinese, American, and Russian Anti-Satellite Testing in Space', *The Space Review*, 17 March 2014, www.thespacereview.com/article/2473/1; Hiroyuki Akita, 'China Ups Ante in Space Arms Race', *Nikkei Asian Review*, 6 January 2015, asia.nikkei.com/Politics/China-ups-ante-in-space-arms-race.

extent of the development and investment in the environment of space by the US—far more than any other nation. As a result, it has developed the greatest dependence upon space of any nation.

However, while the space assets of the US offer extensive services and capabilities, and provide the military with tremendous advantages, these assets are simultaneously virtually defenceless, creating profound vulnerability. This generates a significantly different dynamic than other elements of military power. Such defencelessness is an inherent characteristic of space operations, in that satellites must be light and therefore protection is very difficult, but is compounded by three aspects: the so-called vulnerability gap; nuclear entanglement; and subthreshold, or 'grey zone', tactics.

A vulnerability gap is observed in the advantage that the US enjoys in space relative to other states, allied with the attendant liability, derived from the dependence of the US on these assets. This gap, it can be argued, may create the temptation in certain opponents to attack and disrupt US space capabilities. Such opponents might prosecute an attack in an early phase of a conflict on the assumption of this dependence, hoping to cause surprise and perhaps disarray and a disproportionate advantage.

An opponent might also calculate that if those early attacks on US space assets do not create an existential threat, then there would be minimal risk of retaliation, and the attack would be considered worth the risk. An adversary might also calculate that even if the attack was attributed, and the US responded in a way that destroyed a much greater number or proportion of the enemy's space assets, that adversary would regardless benefit from the action, since overall US military capability would be disproportionately impaired. The result is a strong temptation for asymmetric, pre-emptive attack.

This then intersects with the quandary of nuclear entanglement. Space-based assets have long served as a central component of the US nuclear deterrent, allowing detection of enemy missiles at an early point. For this function, the US uses a small number of satellites. Initially handled by the Milstar constellation, this is transitioning to a small fleet of Advanced Extremely High Frequency (AEHF) satellites. The latter will become the sole space-based system used by the US for 'nuclear communications' once the Milstar satellites are retired. However, the US does not operate discrete communication satellites for nuclear communications. Both Milstar and

AEHF satellites are for nuclear and 'high priority' nonnuclear users, shared by all US military services and some allies. As Wright and colleagues argue, this leads to a high degree of complexity:

> Crucial conventional and nuclear space missions are now deeply entangled, so warfighting with near-peers in space for conventional purposes profoundly threatens the nuclear mission. Commercial and military space systems are also increasingly entangled.[22]

This entanglement is exacerbated by the low number and common type of these assets. The intent of an attack on them may be hard to identify. Combining this with the perceived vulnerability gap increases the likelihood of an attack on US dual-use assets to gain an advantage in a conventional conflict, but by doing so the adversary's nuclear capabilities are affected. The attack may then be judged deliberate, creating the potential for escalation.

A further degree of complexity is presented by subthreshold or grey zone activities, which are those activities that are situated between what might be considered 'normal' competition and what is usually considered to be war. Space is a natural grey zone in many ways—rules about what is permissible in space are vague, and it is very difficult to attribute any action in space to a clear or specific cause. These characteristics make the use of military forces while still avoiding armed clashes less difficult in space than in other environments.

Most familiar are terrestrial examples, such as China utilising 'non-grey hulls' (Coast Guard and fishing vessels) to enhance the state's sovereignty claims in the South China Sea, or Russia's activities in Eastern Ukraine. However, it appears that both of those states have also conducted space operations that correlate with this type of activity. A notable example is China's suspected attack in September 2014 on the National Oceanic and Atmospheric Administration (NOAA) satellite information and weather systems, forcing NOAA to shut the service down for two days.[23] Russia is suspected of manipulating the global positioning system (GPS) to give ships

---

22   Nicholas Wright, ed., *Outer Space; Earthly Escalation? Chinese Perspectives on Space Operations and Escalation. A Strategic Multilayer Assessment*, Periodic Publication (August 2018), 1, nsiteam.com/social/wp-content/uploads/2018/08/SMA-White-Paper_Chinese-Persepectives-on-Space_-Aug-2018.pdf.

23   Micheal Casey, 'China Accused of Hacking into U.S. Weather System', *CBS News*, 12 November 2014, www.cbsnews.com/news/china-accused-of-hacking-into-u-s-weather-system/.

incorrect data, sending them significantly off course in certain instances.[24] Given the broad difficulty of attribution, these attacks are difficult to counter, and partially for this reason, it seems far more likely that the kind of attack that will predominate in the space domain are attacks that dazzle, jam, spoof or manipulate satellites or their signals.[25]

As a result of their dependence on space, many states are sensitive to developments by potential adversaries. This is further amplified by the inherent vulnerability of space assets. This is perceived most keenly by states with the most investment in and dependence on space, particularly the US. With an asymmetry in capabilities, and the ease of prosecuting subthreshold activities, there is potential for insecurity, misperception and thus tension and even conflict in space.

In parallel, the complexion of space is changing, with the commercial sector beginning to dominate this domain. This will in itself shape many aspects of space security, but specifically, it will entwine with military functions in space. They will challenge military capabilities in that they will potentially soon be able to outmatch data collection and processing capabilities, and become the principal source of space-derived information, while also impacting on operational security of military forces. Commercial actors will also develop a number of pioneering technologies and solutions. Although these may be of concern to military forces, these will also increasingly be a source of innovation, and in this respect these new technologies will supplement those developed 'in house' by military programs. But crucially, as commercial interests become entrenched, and large constellations of satellites are put into orbit, these may also help stabilise the space domain. The size and diversity of constellations will add to resiliency against all forms of attacks, kinetic or otherwise, and they will broaden space domain awareness capabilities that can aid transparency and diffuse friction. Even though the rise of the commercial sector will bring challenges in the space domain, it is the potential for aiding transparency and reducing tension that will perhaps be its greatest contribution.

---

24 David Hamblin, 'Ships Fooled in GPS Spoofing Attack Suggest Russian Cyber Weapon', *The New Scientist* 10 August 2017, www.newscientist.com/article/2143499-ships-fooled-in-gps-spoofing-attack-suggest-russian-cyberweapon/; Jim Edwards, 'The Russians are Screwing with the GPS System to Send Bogus Navigation Data to Thousands of Ships', *Business Insider*, 14 April, 2019, www.businessinsider.com/gnss-hacking-spoofing-jamming-russians-screwing-with-gps-2019-4.

25 For an analysis of these threats, see David Livingstone and Patricia Lewis, *Space, the Final Frontier for Cybersecurity?* (Chatham House, International Institute for Strategic Studies, September 2016), www.chathamhouse.org/sites/default/files/publications/research/2016-09-22-space-final-frontier-cyber security-livingstone-lewis.pdf.

References

Casey, Micheal. 'China Accused of Hacking into U.S. Weather System'. *CBS News*, 12 November 2014. www.cbsnews.com/news/china-accused-of-hacking-into-u-s-weather-system/.

Directorate of Intelligence. *Soviet Commercial Space Photography: Offering Resolution as the Solution.* Washington, DC: Directorate of Intelligence, 1990. nsarchive2.gwu.edu/NSAEBB/NSAEBB501/docs/EBB-47.pdf.

Edwards, Jim. 'The Russians are Screwing with the GPS System to Send Bogus Navigation Data to Thousands of Ships'. *Business Insider*, 14 April, 2019. www.businessinsider.com/gnss-hacking-spoofing-jamming-russians-screwing-with-gps-2019-4.

Florini, Ann M, and Yahya A Dehqanzada. *No More Secrets? Policy Implications of Commercial Remote Sensing Satellites.* Carnegie Endowment for International Peace, 1 July 1999. carnegieendowment.org/1999/07/01/no-more-secrets-policy-implications-of-commercial-remote-sensing-satellites-pub-150.

Gruss, Mike. 'Russian Satellite Maneuvers, Silence Worry Intelsat'. *SpaceNews*, 9 October 2015. spacenews.com/russian-satellite-maneuvers-silence-worry-intelsat/.

Gruss, Mike. 'U.S. Intelligence Community Endorses Company's Bid to Sell Sharper Imagery'. *SpaceNews*, 18 April 2014. spacenews.com/40263us-intelligence-community-endorses-companys-bid-to-sell-sharper-imagery/.

Guynn, Jessica. 'Google Buying Skybox Satellite Company to Boost Maps'. *USA Today*, 10 June 2014. eu.usatoday.com/story/tech/2014/06/10/google-buying-skybox-satellite-company/10284045/.

Hamblin, David. 'Ships Fooled in GPS Spoofing Attack Suggest Russian Cyber Weapon'. *The New Scientist,* 10 August 2017. www.newscientist.com/article/2143499-ships-fooled-in-gps-spoofing-attack-suggest-russian-cyberweapon/.

Hiroyuki Akita. 'China Ups Ante in Space Arms Race'. *Nikkei Asian Review*, 6 January 2015. asia.nikkei.com/Politics/China-ups-ante-in-space-arms-race.

Knapp, Alex. 'Forecasting the Weather with Big Data and the Fourth Dimension'. *Forbes*, 13 June 2013. www.forbes.com/sites/alexknapp/2013/06/13/forecasting-the-weather-with-big-data-and-the-fourth-dimension/.

Krepon, Micheal. 'Peacemakers or Rent-a-Spies?' *Bulletin of the Atomic Scientists* 45, no. 7 (September 1989): 12–15. doi.org/10.1080/00963402.1989.11459/16.

Livingstone, David, and Patricia Lewis. *Space, the Final Frontier for Cybersecurity?* Chatham House, International Security Department, September 2016. www.chathamhouse.org/sites/default/files/publications/research/2016-09-22-space-final-frontier-cybersecurity-livingstone-lewis.pdf.

McKinley, Cynthia AS. *When the Enemy Has Our Eyes.* Columbus: Biblioscholar, 2012.

Oremus, Will. 'Google's Eyes in the Sky'. *Slate*, 13 June 2014, slate.com/technology/2014/06/google-skybox-titan-aerospace-acquisitions-why-it-needs-satellites-and-drones.html.

Risen, James, Steven Lee Myers and Tim Weiner. 'U.S. May Have Helped India Hide Its Nuclear Activity'. *New York Times,* 25 May 1998. www.nytimes.com/1998/05/25/world/us-may-have-helped-india-hide-its-nuclear-activity.html.

Samuels, David. 'Inside a Startup's Plan to Turn a Swarm of DIY Satellites into an All-Seeing Eye'. *Wired*, 18 June 2013. www.wired.com/2013/06/startup-skybox/.

'SkySat Constellation'. Satellite Mission Catalogue, *eoPortal.* directory.eoportal.org/web/eoportal/satellite-missions/s/skysat.

Space Foundation. 'Commercial Space Revenue Climbs to All-Time High of $328.86 Billion, the Space Report Reveals in Q3 Analysis'. *Space Foundation*, 31 October 2019. www.spacefoundation.org/2019/10/31/commercial-space-revenue-climbs-to-all-time-high-of-328-86-billion-the-space-report-reveals-in-q3-analysis/.

Union of Concerned Scientists. *UCS Satellite Database*, 8 December 2005. www.ucsusa.org/resources/satellite-database.

US Army Headquarters. *Field Manual 100-5, Operations, May 1986.* Washington, DC: Department of the Army, 1986.

'US Lifts Restrictions on More Detailed Satellite Images'. *BBC News*, 16 June 2014. www.bbc.co.uk/news/technology-27868703.

Weeden, Brian. 'Through a Glass, Darkly: Chinese, American, and Russian Anti-Satellite Testing in Space'. *The Space Review*, 17 March 2014. www.thespacereview.com/article/2473/1.

Weeden, Brian, and Victoria Samson, eds. *Global Counterspace Capabilities: An Open Source Assessment.* Washington, DC: Secure World Foundation, April 2018. swfound.org/media/206118/swf_global_counterspace_april2018.pdf.

Wright, Nicholas, ed. *Outer Space; Earthly Escalation? Chinese Perspectives on Space Operations and Escalation. A Strategic Multilayer Assessment*. Periodic Publication. August 2018. nsiteam.com/social/wp-content/uploads/2018/08/SMA-White-Paper_Chinese-Persepectives-on-Space_-Aug-2018.pdf.

Zimmerman, Peter D. 'Introduction to Photo-Interpretation of Commercial Observation-Satellite Imagery'. In *Commercial Observation Satellites and International Security*, edited by Michael Krepon. Basingstoke: Macmillan in association with the Carnegie Endowment for International Peace, 1990.

5

Space Warfare and Military Power

Malcolm Davis

Introduction[1]

Space is a warfighting domain.[2] Although it is a global common—much like the oceans, or cyberspace—it isn't a sanctuary that sits placid and serene, untouched by terrestrial geopolitical rivalry and conflict below. Space has been militarised since the dawn of the space age, with satellites used to support a broad range of terrestrial military tasks, notably nuclear command and control—a reality from the early 1960s. Over the decades, the role of space has spread into other military tasks, with more ubiquitous satellite communications, advanced intelligence surveillance and reconnaissance capabilities, and, from the 1980s, global navigation satellite systems such as the US GPS.

As the technology of space has matured, the dependency of military forces on space capabilities has expanded, and more states are investing in such capabilities as force multipliers. The growing dependency of terrestrial military forces on space support has given incentive towards developing

---

1   Editors' note: as mentioned in the Introduction, this chapter was written in 2020, and global events since then may have overtaken some aspects. As the pace of technological and cultural change has continued to accelerate, the editors have opted to present this text as drafted to minimise further delays in publication.

2   This refers to the capability of denying space to adversaries and differs from definitions of space as an operational domain, which are related to the integration and interoperability of space assets with other military domains, with space assets enabling wider military operations.

counterspace capabilities. These include 'hit to kill' anti-satellite (ASATs) weapons and 'soft kill' systems designed to disable rather than destroy a target satellite.

ASAT weapons are not new, and both the US and the Soviet Union were developing different types of ASAT capabilities throughout the Cold War.[3] However, the US never operationally deployed such capability, and the Soviet Union only undertook a limited deployment of its ASAT systems in the 1970s. The winding down of Cold War tensions in the late 1980s, and the opportunity of arms control, meant that counterspace development remained largely moribund through to the early 21st century.

China's test of an ASAT on 11 January 2007 fundamentally changed the debate over counterspace and brought the importance of such systems back to the forefront. It overturned assumptions in the West that space was accepted by all as a peaceful commons and challenged ideas that states would have unchallenged access to space support in warfare.[4] With accelerating counterspace programs now emerging in China, Russia and the US, including the operational deployment of actual ASAT weapons by China, and with the development of more sophisticated co-orbital and 'soft kill' counterspace capability underway, space is now moving from being militarised towards becoming weaponised.[5] It raises the growing prospect of space warfare occurring in future terrestrial conflicts, either spreading from Earth into space, or an attack on space systems generating conflict on Earth.

Towards space warfare

Australian defence planners have started to come to terms with the challenges posed by counterspace capabilities and the risk of space warfare denying us access to critical space support in wartime. The potential risks of counterspace capability were identified by the Australian defence policy community in the 2016 Defence White Paper, which stated:[6]

---

3   Brian Weeden, *Through a Glass, Darkly: Chinese, American and Russian Anti-Satellite Testing in Space* (Secure World Foundation, 17 March 2014), swfound.org/media/167224/through_a_glass_darkly_march2014.pdf.

4   Brian Weeden, *2007 Chinese Anti-Satellite Test Fact Sheet* (Secure World Foundation, 2010), swfound.org/media/9550/chinese_asat_fact_sheet_updated_2012.pdf.

5   Todd Harrison, Kaitlyn Johnson, Thomas G Roberts, with Madison Bergethon and Alexandra Coultup, *Space Threat Assessment 2019*, Report of the CSIS Aerospace Security Project (Washington, DC: Center for Strategic and International Studies, April 2019).

6   Australian Department of Defence, *2016 Defence White Paper* (Canberra: Australian Department of Defence, 2016), 52 (2.53), 87 (4.16).

> Some countries are developing capabilities to target satellites to destroy these systems or degrade their capabilities, threatening our networks.

and then went on to note that:

> Satellite systems are vulnerable to space debris, which could damage or disable satellites, and advanced counter-space capabilities such as anti-satellite missiles, which can deny, disrupt and destroy our space-based systems.

Since then, there has been broad policy consensus within Australia's defence policy community, and also shared by our key allies, that the 21st Century space domain is 'contested, congested and competitive'.[7] Space is 'contested' as counterspace capabilities including 'soft kill' measures emerge in the military forces of potential future adversaries. Space is 'congested' as a result of the rapid increase in man-made objects such as satellites and the growth of space debris, which act to complicate space operations and increase the risk of collisions in orbit. The risk of use of some ASAT systems adds to the potential for congestion. Space is 'competitive' due to the increasing number of space actors—both nation-state and commercial—which are enjoying greater access to space at lower cost largely due to the Space 2.0 transformation.[8] These trends are only set to continue.

Western military forces are highly dependent on space systems to undertake military operations in an effective manner. They could not fight an information-based war that is fast, precise and minimises the risk of casualties in terms of its forces, or the risk to civilian populations, if they lost access to space systems. Instead, they would be forced to revert to a more basic industrial approach, which would see higher casualties, prolonged hostilities and a greater risk of defeat.

Understanding this new high frontier of warfare in the 21st century is vital to grasp how future warfare might occur. The next major power war is likely to start in space (and also, in cyberspace—indeed the two domains increasingly overlap) as our adversaries seek to deny us access to critical

---

7   Malcolm Davis, 'Australia Takes on the High Frontier', *The Strategist* (blog), *Australian Strategic Policy Institute*, 2 March 2018, www.aspistrategist.org.au/australia-takes-high-frontier.

8   Malcolm Davis, *Australia's Future in Space, Strategy* (Canberra: Australian Strategic Policy Institute, February 2018).

space-based systems for 'C4ISR' and 'PNT'.[9] This could happen prior to, or at the outset of a military conflict, with the goal of future adversaries being to prevent the ability of the US, and its allies, including Australia, from gaining and sustaining a critical knowledge edge that would translate to a significant military advantage on the battlefield. Taking away that knowledge edge certainly erodes those advantages, reduces our ability to wage information-based joint warfare, and levels the field such that some adversaries with significant counterspace capability, such as China and Russia, are better placed to exploit mass and long-range firepower to our disadvantage.

ASATs and counterspace

The broad capability area of 'counterspace' encompasses a range of systems and technologies. The most basic (and the most commonly tested) are 'hit to kill' ASATs that are launched from Earth's surface, or from the air, against target satellites in orbit. Such systems—known as 'direct-ascent ASATs' or 'DA-ASTs'—were developed during the Cold War by the former Soviet Union and the US, but only the Soviets operationally deployed such weapons on a limited basis. China's test of a DA-ASAT on 11 January 2007, noted above, was the first since the end of the Cold War and transformed the debate in the West on space security, shattering the myth that space was a peaceful commons. India most recently tested a DA-ASAT on 27 March 2019, destroying a target satellite in low-Earth orbit (LEO).[10] The US did an unofficial ASAT test on 20 February 2008 during Operation Burnt Frost, when it used a sea-based SM-3 interceptor missile to strike a malfunctioning satellite and prevent a potentially hazardous uncontrolled re-entry.[11] China and Russia continue to develop direct-ascent ASATs, and US Defense Intelligence Agency analysis suggests that China has likely operationally deployed such weapons since 2010.[12] China also tested

---

9   'C4ISR' is 'Command, Control, Communications, Computers, Intelligence, Surveillance and Reconnaissance'. 'PNT' is 'Position, Navigation and Timing'.

10   Malcolm Davis, 'Will India's Anti-Satellite Weapon Test Spark an Arms Race in Space?' *The Strategist* (blog), *Australian Strategic Policy Institute*, 29 March 2019, www.aspistrategist.org.au/will-indias-anti-satellite-weapon-test-spark-an-arms-race-in-space/.

11   Nicole Petrucci, 'Reflections on Operation Burnt Frost', *Air Power Strategy*, 5 March 2017, www.airpowerstrategy.com/2017/03/05/burnt-frost/.

12   Office of the Director of National Intelligence, *Annual Threat Assessment of the U.S. Intelligence Community* (Washington, DC: Defense Intelligence Agency, February 2022), 8, www.odni.gov/files/ODNI/documents/assessments/ATA-2022-Unclassified-Report.pdf.

a high-Earth orbit delivery system on 13 May 2013 that could be employed to deliver an ASAT into geostationary orbit, which is the location of US and allied communication satellites, as well as cover satellites in medium-Earth orbit including US GPS and missile early warning sensors such as the Space-based Infra-red System (known as SBIRS) satellites.[13]

A second type of anti-satellite weapon is the 'co-orbital ASAT'. As the name implies, this is a weapon that is placed into orbit, and slowly approaches its target in a 'rendezvous and proximity operation' (RPO) rather than rapidly targeting and destroying a satellite from the surface or within the atmosphere. Co-orbital ASATs can use 'hit to kill', but they can also employ 'soft kill' mechanisms at close range, such as electronic warfare including close-range jamming, or physical interference via a robotic arm.

'Soft kill' offers a number of advantages, notably, they are designed to disable or disrupt, rather than physically destroy. 'Soft kill' mechanisms can generate scalable and reversible effects, and they don't generate large clouds of space debris, unlike 'hit to kill' ASATs. They offer a would-be user of ASATs the chance to generate grey zone operations in orbit because the same principles for their use in ASAT roles can also be applied to non-military, non-hostile actions in terms of commercial on-orbit refuelling and repair of satellites. An adversary can thus develop and deploy a co-orbital ASAT capability covertly, hidden behind civil commercial space capability.

A state could reasonably declare that it has an interest in pursuing a commercial on-orbit refuelling and repair capability as part of the civil side of its space program. This would be a legitimate business exploiting a new technological means to extend and renew satellites in orbit and upgrade or repair them when necessary. Certainly, on-orbit repair and refuel is emerging as a potentially lucrative part of the commercial space sector globally. At the same time, this would allow a state to test the technologies of a co-orbital ASAT capability under a commercial guise, employing the same RPO close approach and docking manoeuvres as would a civilian or commercial activity. Prior to, or at the outset of a conflict, China could then rapidly transition such a commercial capability to a military role as

---

13   Brian Weeden and Victoria Sampson, eds, *Global Counterspace Capabilities: An Open Source Assessment* (Washington, DC: Secure World Foundation, April 2019), 1-11, swfound.org/media/206408/swf_global_counterspace_april2019_web.pdf.

co-orbital ASATs. This would be consistent with language from the People's Liberation Army (PLA) Academy of Military Science, which in a recent report on space warfare argued that China would:

> strive to attack first at the campaign and tactical levels in order to maintain the space battlefield initiative, and that the intent should be to conceal concentration of forces and make a decisive large-scale first strike.[14]

This approach—rather than reliance only on DA-ASATs designed for the physical destruction of a target (generating space debris)—seems a more likely path for future adversary counterspace campaigns. Chinese researchers are also discussing other approaches such as using artificial intelligence (AI)–driven small satellites to 'capture' adversary satellites.[15] Such measures expand the spectrum of potential counterspace threats facing the US and its allies, including Australia, over the coming decade and beyond.

Reinforcing the grey zone phenomenon risk, ground-based counterspace technologies add potential degrees of anonymity and deniability for counterspace operations that a state could exploit prior to, or at the outset of, a military conflict. China's PLA Strategic Support Force (PLASSF) covers both space and network operations, including electronic warfare operations. A recent RAND report on the role of the PLASSF argues that China's approach to counterspace—'space attack and defense operations'—seeks to achieve superiority within a certain period of time and within a certain location, and ground-based counterspace would form a 'third leg' of China's capability for space attack and defence.[16]

A key 2016 Chatham House report argued that satellites are vulnerable to cyberattack, both directly and indirectly through ground stations, or through the supply chain of satellite technology.[17] Although military satellites are often hardened against cyber intrusion, commercial satellites

---

14   Kevin Pollpeter, *Testimony before the US-China Economic and Security Review Commission Hearing on 'China in Space: Strategic Competition'* (Washington, DC: CNA Analysis and Solutions, 25 April 2019), 6; see also Jiang Lianju and Wang Liwen, eds, *Textbook for the Study of Space Operations* (Beijing: Military Science Publishing House, 2013), 44.
15   See Stephen Chen, 'China Develops AI That "Can Use Deception to Hunt Satellites"', *South China Morning Post*, 13 June 2022, www.scmp.com/news/china/science/article/3181546/china-develops-ai-can-use-deception-hunt-satellites.
16   Pollpeter, *Testimony before the US-China Economic and Security Review Commission Hearing*, 9.
17   David Livingstone and Patricia Lewis, *Space, the Final Frontier for Cybersecurity?* (Chatham House, International Security Department, September 2016), www.chathamhouse.org/sites/default/files/publications/research/2016-09-22-space-final-frontier-cybersecurity-livingstone-lewis.pdf.

used by military forces often are not. The very character of cyberwarfare and computer network operations means that states can exploit this form of power without necessarily incriminating themselves, and they can do so well before a declaration of war or an outbreak of overt military hostilities, and the nature of cyber threats allows them to be insidiously planted inside critical systems in a dormant state. Cyberattacks on satellites might begin not on the day of war breaking out, but months or even years beforehand when malicious code is slipped into critical components through the commercial supply chain and carried out by undeclared cyber forces acting on behalf of a state.

The Chatham House report lays out a number of potential cyberattack methods:[18]

a. jamming, spoofing and hacking attacks on communication networks via space infrastructure

b. attacks on satellites, targeting their control systems or mission packages, perhaps taking control of a satellite to exploit its capabilities, shut it down, alter its orbit or 'cook' or 'grill' its solar cells through deliberate exposure to damaging levels of radiation

c. attacks on ground infrastructure, such as satellite control centres, associated networks and data centres, leading to potential global cascading effects on critical information infrastructure and networks.

The nature of cyberwarfare suggests some important risks for maintaining access and freedom of action in space.

Firstly, the potential for this type of counterspace attack is open to a much broader range of international actors, including non-state actors such as international terrorist networks like Al Qaeda and Islamic State, and potentially even the would-be hacker seeking a new challenge. It is not a challenge that is relevant only to high-intensity interstate warfare scenarios. This means Western states will need to think about, and plan for, how cyberattacks on satellites can be a factor in a full range of lower-intensity scenarios.

Secondly, cyberattacks, together with uplink and downlink jamming, and laser dazzling, allow scalable and reversible effects, with cyberattacks also offering an intelligence-gathering role through the monitoring or theft of

---

18   Livingstone and Lewis, *Space*, 9–10.

information passing through a satellite or via a ground station. Cyberattacks can also be employed to 'spoof' a satellite to generate false information to an opponent in a deception campaign. The Russians have already tried this several times, including against US GPS satellites during recent NATO (North Atlantic Treaty Organization) exercises off northern Norway, and also interfering with GPS in a manner that misdirected commercial shipping in the Black Sea.[19] The potential for coercion, hostile intelligence operations, and strategic information warfare through cyber operations against a state's space segment must be a real concern.

Thirdly, a key advantage of cyberattacks as a means to undertake counterspace operations is their low cost. A cyberattack can be quickly undertaken through a broader offensive and defensive computer network operations program, open to non-state actors that only need computers and a good understanding of hacking to develop capabilities that could be effective, at least against non-hardened commercial satellites. The footprint of cyber counterspace activities is minimal compared to traditional counterspace capabilities, such as DA-ASATs. There's no need for rocket technology, advanced spacecraft technology, or highly visible testing with the equally visible and politically damaging debris clouds that such tests generate. Detecting the development of such capability therefore is highly challenging for intelligence organisations, and there is little or no possibility of preventing the development of such a capability if a state—or non-state actor—chooses to acquire it.

Fourthly, cyberattacks on satellites can more readily leverage advances in the civil sector, including rapid developments in AI to develop highly potent counterspace capability. The potential implications of AI in controlling cyber operations, including for counterspace, are largely conjecture at this time, but given the benefits of machine speed analysis, command and control, it is likely that an AI-directed counterspace campaign employing cyber capabilities would be quicker, more efficient and less likely to fail than a human-directed campaign.

---

19   C4ADS, *Above Us Only Stars: Exposing GPS Spoofing in Russia and Syria* (C4ADS, 2019) c4ads.org/reports/above-us-only-stars/; Michael Jones, 'Spoofing in the Black Sea: What Really Happened?' *GPS World*, 11 October 2017, www.gpsworld.com/spoofing-in-the-black-sea-what-really-happened/.

Is warfare in space inevitable?

Clearly, the means to undertake space warfare, in terms of military technologies and systems, are either here and potentially in operational service, or rapidly being developed. Peer adversaries such as China and Russia, while promoting declaratory policies embracing norms of non-weaponisation of space, are busy rapidly developing a full suite of counterspace capabilities.[20] The US, having emphasised a strong degree of strategic restraint on space weaponisation during the Obama Administration, is now moving to protect its vital access to space and established an independent space force in December 2019.

The US certainly has the option to develop its own counterspace capability. Its ballistic missile defence capabilities have dual-role ASAT potential and could be transitioned into an operational DA-ASAT weapon against LEO-based satellites. The US has also demonstrated co-orbital RPOs, like China and Russia, for satellite inspection and intelligence-gathering purposes. If it chose to do so, the US could develop a range of ground-based 'soft kill' systems including cyberattack.

Current US counterspace doctrine emphasises the importance of achieving space superiority which is:

> The degree of control in space of one force over any others that permits the conduct of its operations at a given time and place without prohibitive interference from terrestrial or space-based threats.[21]

US counterspace operations are divided into defensive and offensive counterspace concepts, with the latter encompassing the full range of measures including deceiving, disrupting, denying, degrading or, if necessary, destroying adversary space capabilities, while defensive counterspace seeks to protect US and friendly capability from attack and reinforce credible space deterrence.

---

20 Weeden and Sampson, *Global Counterspace Capabilities*.
21 United States Air Force, *Counterspace Operations*, Air Force Doctrine Publication 3-14 (Curtis E LeMay Center for Doctrine Development and Education, August 2018).

Space deterrence is a key concept emerging in Western space policy thinking, in which the objective is to change the calculus of a peer adversary with counterspace capabilities such that the adversary chooses not to use such capabilities.[22] A key component is emphasising the development of space resilience by reducing the prospects of a successful adversary counterspace campaign. If US and allied space capability can be resilient against an opponent's ASATs and counterspace systems, the opponent may calculate that the benefits gained in using counterspace weapons are outweighed by the potential risks and costs in doing so.

Strengthening US and allied space resilience can be achieved through augmentation, disaggregation and reconstitution. A key vulnerability of US and allied space systems is the reliance on small numbers of large, complex and expensive satellites for satellite communications, intelligence, surveillance and reconnaissance, and precision navigation and timing. Providing larger numbers of smaller satellites prior to a conflict complicates an adversary's ability to use counterspace capability. Smaller lower-cost satellites, including CubeSats, can be low cost, deployed quickly into orbit, and can support disaggregation of constellations, to share tasks across multiple satellites.[23] That further complicates an adversary's task, by switching to the 'small and many' operational approach and making it harder for an adversary to undertake a decisive counterspace campaign. Finally, if an adversary does use counterspace systems, the rapid reconstitution of space capabilities through the rapid launch of small satellites to restore space systems avoids catastrophic loss of space support.

Critical in this regard is space situational awareness (SSA), which is a task that Australia is heavily involved with alongside the US and other allies through the Australia–US Space Situational Awareness Partnership, and through initiatives such as the Common Space Operations Initiative.[24] Australia's current Defence White Paper, released in 2016, highlights the SSA role, noting:

---

22 Todd Harrison, Kaitlyn Johnson, and Thomas G Roberts, *Escalation and Deterrence in the Second Space Age* (Washington, DC: Center for Strategic and International Studies, 3 October 2017), www.csis.org/analysis/escalation-and-deterrence-second-space-age.

23 Davis, *Australia's Future in Space.*

24 Cheryl Pellerin, 'Stratcom, DoD Sign Space Operations Agreement with Allies', *US Department of Defense*, 23 September 2014, dod.defense.gov/News/Article/Article/603303/stratcom-dod-sign-space-operations-agreement-with-allies/.

In cooperation with the United States, Australia is strengthening its space surveillance and situational awareness capabilities. At the centre of this work is the establishment of the space surveillance C-band radar operated jointly by Australia and the United States, and the relocation of a United States optical space surveillance telescope to Australia. The radar and telescope will increase our capacity to detect and track objects in space, including space debris, and predict and avoid potential collisions.[25]

SSA is crucial to countering adversary counterspace activities, denying states engaging in hostile actions anonymity, and reducing the possibility of success of grey zone activities in orbit. Ensuring attribution of counterspace threats is a key step towards strengthening space deterrence and allows greater space resilience, including the potential for defensive counterspace activities such as manoeuvring satellites under threat. It's an integral part of Australia–US cooperation in space.

Space deterrence thus is primarily based on deterrence by denial through enhancing space resilience, together with clear communication to an opponent of costs imposed in the event an adversary does employ counterspace capability. It doesn't make space warfare impossible, but space deterrence *does* reduce the risk of it happening, by making the use of counterspace capabilities less appealing to a would-be space adversary.

Space deterrence and space law

Space deterrence also can contribute towards creating the right conditions for establishing legal constraints on space weaponisation and restoring norms against space weapons that have been seriously eroded, particularly as a result of Chinese and Russian counterspace capability development. By making the use of counterspace capability less likely to generate a useful military effect, together with increasing potential costs of doing so, there is a greater incentive for all sides to establish legal norms that would then open up prospects for verifiable arms control of counterspace capability.

---

25 Australian Department of Defence, *2016 Defence White Paper*, 4.16.

Current space law, centred around the 1968 Outer Space Treaty (OST), does not prohibit the development or deployment of weapons in space, with the exception of weapons of mass destruction.[26] Nor does it address the potential for a rapid expansion of commercial space activity to potentially circumvent the key principles of the OST for claiming territory on the Moon and other celestial bodies, which could, in turn, weaken the OST and increase the risk of major power 'astropolitical' competition in the future.

There is ongoing dialogue through international legal frameworks such as the UN Conference on Disarmament and the Committee on the Peaceful Uses of Outer Space (COPUOS), together with Track 2 efforts such as the Woomera Manual and the MILAMOS Project, which are providing international policy and legal dialogue frameworks that could directly complement the positive effects of space deterrence and strengthen legal constraints on space weaponisation.[27] This could include the formulation of complementary legal agreements to the OST that fill the shortcomings of that treaty in terms of space weaponisation. Such dialogue also could contribute to what Annie Handmer refers to as strategic space diplomacy, which would build dialogue between all sides and potentially contribute to an easing of military competition in orbit.[28]

However, the key challenge in relying on legal approaches to constrain or ban space weapons work is the huge difficulty in verification and enforcement, which relates to the question of 'what defines a space weapon' when even a low-cost CubeSat, if redirected to collide with another satellite, can act as a co-orbital ASAT.[29] The impact of dual-role space capabilities, and the intangible aspects of ground-based approaches, such as cyberattacks on satellites, makes legal means towards banning counterspace weapons (with the exception of traditional DA-ASAT systems) virtually impossible to enforce. Co-orbital ASATs use the same RPOs as commercial on-orbit

---

26 Malcolm Davis, 'Avoiding a Free-for-All: The Outer Space Treaty Revisited', *The Strategist* (blog), *Australian Strategic Policy Institute*, 16 July 2018, www.aspistrategist.org.au/avoiding-a-free-for-all-the-outer-space-treaty-revisited/.

27 United Nations Office for Outer Space Affairs, 'Committee on the Peaceful Uses of Outer Space', *UN Office for Outer Space Affairs*, www.unoosa.org/oosa/en/ourwork/copuos/index.html; MILAMOS refers to the *Manual on International Law Applicable to Military Uses of Outer Space*, published by McGill University, www.mcgill.ca/milamos/; The University of Adelaide, The Woomera Manual on the International Law of Military Space Activities and Operations project, law.adelaide.edu.au/woomera/home.

28 Annie Handmer, 'Australia Should Aspire to Be a Leader in Strategic Space Diplomacy', *The Strategist* (blog), *Australian Strategic Policy Institute*, 31 October 2018, www.aspistrategist.org.au/australia-should-aspire-to-be-a-leader-in-strategic-space-diplomacy/.

29 Malcolm Davis, 'Space 2.0—Enabling War in Space?' *The Strategist* (blog), *Australian Strategic Policy Institute*, 9 May 2019, at www.aspistrategist.org.au/space-2-0-enabling-war-in-space/.

refuel and repair systems—a classic dual-role capability which if banned would also by definition leave a key part of the 21st-century commercial space sector stillborn.

Efforts through initiatives such as Prevention of an Arms Race in Outer Space (PAROS) and the Prohibition on Placement of Weapons Treaty (PPWT), sponsored by Russia and China in international bodies such as COPUOS and the UN Conference on Disarmament, have failed to gain broad international support because they don't address the issue of verification and monitoring effectively, and would leave existing Russian and Chinese ASAT capabilities intact while preventing a balancing counterspace capability being developed by the US.[30]

This is not to suggest that efforts towards legal constraints on space weapons, strengthening space law and enhancing regulation against destabilising technologies are a waste of time. A key task for space law should be banning 'hit to kill' ASAT technologies that create space debris. The risk of rapid use of such systems against an opponent's satellites is that space would rapidly fill with clouds of space debris that would choke critical orbits and trajectories, denying access to space for all parties. The potential for such an event to generate cascading collisions—known as 'Kessler Syndrome'— making space inaccessible for generations, is real, and widespread use of kinetic kill ASATs would make such an event much more likely.[31] If there is a key objective for legal efforts towards preventing space weaponisation, it must be first directed at establishing internationally accepted principles and norms against such 'hit to kill' systems, with appropriate verification and monitoring measures in place.

Conclusions

Future wars will be fought across all operational domains, including in space. The emerging nature of space warfare raises the prospect of a 'Pearl Harbour in space', where peer adversary states launch a lightning attack on vital US

---

30 Nuclear Threat Initiative, 'Proposed Prevention of an Arms Race in Space (PAROS) Treaty', *Nuclear Threat Initiative*, 29 September 2017, at www.nti.org/learn/treaties-and-regimes/proposed-prevention-arms-race-space-paros-treaty/; Michael Listner and Rajeswari Rajagopalan, 'The 2014 PPWT: A New Draft but with the Same and Different Problems', *The Space Review*, 11 August 2014, www.thespacereview.com/article/2575/1.

31 Scott Kerr, 'Liability for Space Debris Collisions and the Kessler Syndrome (Part 1)', *The Space Review*, 11 December 2017, www.thespacereview.com/article/3387/1.

and allied space-based C4ISR and PNT architecture to deny access to these critical space systems, leaving terrestrial forces deaf, dumb and blind. The use of US and allied space capabilities to gain and maintain a knowledge edge over an opponent is both a strength—space gives an unparalleled global vantage from which to understand the battlespace and assure rapid and effective command and control of air, sea and land forces—and a weakness due to our acute dependency on this new domain. In this sense, the space domain is emerging as a new centre of gravity for an opponent to concentrate firepower in both time and space to produce rapid and decisive strategic effects. The US and its allies, including Australia, are very cognisant of the risks that dependency on space capabilities raises and are seeking to ameliorate those risks by boosting space deterrence, emphasising resilience and enhancing space surveillance. New organisational structures are emerging to cope with a space domain that is contested, congested and competitive, and—so far—the emphasis is more on defensive counterspace and resilience, rather than US development and deployment of offensive counterspace capabilities such as ASATs. Certainly, legal measures, space diplomacy and arms control may contribute to easing some of the risks posed by adversary counterspace capability, and efforts should be directed at reducing the challenge of space debris as a priority.

The reality that we must confront is that the era of space warfare is upon us. Space is no longer a peaceful commons, but is highly militarised and likely to become weaponised. In a new era of major power competition, the risks of the use of space weapons are rising. We must learn from Sir Bernard Montgomery of El Alamein who referred to the importance of gaining and sustaining control of the air, noting: 'If we lose the war in the air, we lose the war and we lose it quickly.'[32] The same is true for the importance of ensuring and maintaining access to space.

References

Australian Department of Defence. *2016 Defence White Paper.* Canberra: Australian Department of Defence, 2016.

C4ADS. *Above Us Only Stars: Exposing GPS Spoofing in Russia and Syria.* C4ADS, 2019. c4ads.org/reports/above-us-only-stars/.

---

32   Joint Chiefs of Staff, *Countering Air and Missile Threats*, Joint Publication 3-01 (United States: US Department of Defense, 21 April 2017), I-1, at fas.org/irp/doddir/dod/jp3_01.pdf.

Chen, Stephen. 'China Develops AI That "Can Use Deception to Hunt Satellites"'. *South China Morning Post*, 13 June 2022. www.scmp.com/news/china/science/article/3181546/china-develops-ai-can-use-deception-hunt-satellites.

Davis, Malcolm. 'Australia Takes on the High Frontier'. *The Strategist* (blog), *Australian Strategic Policy Institute*, 2 March 2018. www.aspistrategist.org.au/australia-takes-high-frontier.

Davis, Malcolm. *Australia's Future in Space. Strategy*. Canberra: Australian Strategic Policy Institute, February 2018.

Davis, Malcolm. 'Avoiding a Free-for-All: The Outer Space Treaty Revisited'. *The Strategist* (blog), *Australian Strategic Policy Institute*, 16 July 2018. www.aspistrategist.org.au/avoiding-a-free-for-all-the-outer-space-treaty-revisited/.

Davis, Malcolm. 'Space 2.0—Enabling War in Space?' *The Strategist* (blog), *Australian Strategic Policy Institute*, 9 May 2019. www.aspistrategist.org.au/space-2-0-enabling-war-in-space/.

Davis, Malcolm. 'Will India's Anti-Satellite Weapon Test Spark an Arms Race in Space?' *The Strategist* (blog), *Australian Strategic Policy Institute*, 29 March 2019. www.aspistrategist.org.au/will-indias-anti-satellite-weapon-test-spark-an-arms-race-in-space/.

Handmer, Annie. 'Australia Should Aspire to Be a Leader in Strategic Space Diplomacy'. *The Strategist* (blog), *Australian Strategic Policy Institute*, 31 October 2018. www.aspistrategist.org.au/australia-should-aspire-to-be-a-leader-in-strategic-space-diplomacy/.

Harrison, Todd, Kaitlyn Johnson, and Thomas G Roberts. *Escalation and Deterrence in the Second Space Age*. Washington, DC: Center for Strategic and International Studies, 3 October 2017. www.csis.org/analysis/escalation-and-deterrence-second-space-age.

Harrison, Todd, Kaitlyn Johnson, Thomas G Roberts, with Madison Bergethon and Alexandra Coultup. *Space Threat Assessment 2019*. Report of the CSIS Aerospace Security Project. Washington, DC: Center for Strategic and International Studies, April 2019.

Jakhu, Ram S, and Steven Freeland. *McGill University Manual on International Law Applicable to Military Uses of Outer Space*. Montreal: McGill University, 2022. www.mcgill.ca/milamos/.

Jiang Lianju, and Wang Liwen, eds. *Textbook for the Study of Space Operations*. Beijing: Military Science Publishing House, 2013.

Joint Chiefs of Staff. *Countering Air and Missile Threats*, Joint Publication 3-01. United States: US Department of Defense, 21 April 2017.

Jones, Michael. 'Spoofing in the Black Sea: What Really Happened?' *GPS World*, 11 October 2017. www.gpsworld.com/spoofing-in-the-black-sea-what-really-happened/.

Kerr, Scott. 'Liability for Space Debris Collisions and the Kessler Syndrome (Part 1)'. *The Space Review*, 11 December 2017. www.thespacereview.com/article/3387/1.

Listner, Michael, and Rajeswari Rajagopalan. 'The 2014 PPWT: A New Draft but with the Same and Different Problems'. *The Space Review*, 11 August 2014. www.thespacereview.com/article/2575/1.

Livingstone, David, and Patricia Lewis. *Space, the Final Frontier for Cybersecurity?* Chatham House, International Security Department, September 2016. www.chathamhouse.org/sites/default/files/publications/research/2016-09-22-space-final-frontier-cybersecurity-livingstone-lewis.pdf.

Nuclear Threat Initiative. 'Proposed Prevention of an Arms Race in Space (PAROS) Treaty'. *Nuclear Threat Initiative*, 29 September 2017. www.nti.org/learn/treaties-and-regimes/proposed-prevention-arms-race-space-paros-treaty/.

Office of the Director of National Intelligence. *Annual Threat Assessment of the U.S. Intelligence Community.* Washington, DC: Defense Intelligence Agency, 8 March 2022. www.odni.gov/files/ODNI/documents/assessments/ATA-2022-Unclassified-Report.pdf.

Pellerin, Cheryl. 'Stratcom, DoD Sign Space Operations Agreement with Allies'. *US Department of Defense*, 23 September 2014. dod.defense.gov/News/Article/Article/603303/stratcom-dod-sign-space-operations-agreement-with-allies/.

Petrucci, Nicole. 'Reflections on Operation Burnt Frost'. *Air Power Strategy*, 5 March 2017. www.airpowerstrategy.com/2017/03/05/burnt-frost/.

Pollpeter, Kevin. *Testimony before the US-China Economic and Security Review Commission Hearing on 'China in Space: Strategic Competition'.* Washington, DC: CNA Analysis and Solutions, 25 April 2019.

United Nations Office for Outer Space Affairs. 'Committee on the Peaceful Uses of Outer Space'. *UN Office for Outer Space Affairs.* www.unoosa.org/oosa/en/ourwork/copuos/index.html.

United States Air Force. *Counterspace Operations.* Air Force Doctrine Publication 3-14. Curtis E LeMay Center for Doctrine Development and Education, August 2018.

Weeden, Brian. *2007 Chinese Anti-Satellite Test Fact Sheet.* Secure World Foundation, 2010. swfound.org/media/9550/chinese_asat_fact_sheet_updated_2012.pdf.

Weeden, Brian. *Through a Glass, Darkly: Chinese, American and Russian Anti-Satellite Testing in Space.* Secure World Foundation, 17 March 2014. swfound.org/media/167224/through_a_glass_darkly_march2014.pdf.

Weeden, Brian, and Victoria Sampson, eds. *Global Counterspace Capabilities: An Open Source Assessment.* Washington, DC: Secure World Foundation, April 2019. swfound.org/media/206408/swf_global_counterspace_april2019_web.pdf.

6

China and Space Warfare

Jian Zhang

Introduction[1]

China is rapidly emerging as a leading power in outer space. Over the last two decades or so, the country has made striking progress in its space program, including six successful manned orbit missions with a spacewalk and the launch of a habitable space station program. In January 2019, China successfully landed a rover, Chang'e 4, on the far side of the Moon, becoming the first country in the world to do so. Chinese astronauts could also land on the Moon in the next decade.[2] China currently has the second-largest number of operational satellites after the US and its space fleet is growing rapidly.[3] In 2018 and 2019, China undertook more space launches than any other country in the world.[4] China's 2016 space white paper, entitled *China's Space Activities in 2016*, stated explicitly Beijing's intention to become a 'space power [航天强国] in all aspects'.[5]

---

1   Editors' note: as mentioned in the Introduction, this chapter was written in 2020, and global events since then may have overtaken some aspects. As the pace of technological and cultural change has continued to accelerate, the editors have opted to present this text as drafted to minimise further delays in publication.

2   Alexander Bowe, *China's Pursuit of Space Power Status and Implications for the United States*, Staff Research report (Washington, DC: U.S.–China Economic and Security Review Commission, 11 April 2019).

3   China Power, 'How Is China Advancing Its Space Launch Capabilities?', *China Power*, 5 November 2019, chinapower.csis.org/china-space-launch/.

4   Andrew Jones, 'China to Continue World-Leading Launch Rate in 2020', *Space News*, 3 December 2019, spacenews.com/china-to-continue-world-leading-launch-rate-in-2020/.

5   *China's Space Activities in 2016* (white paper), (Beijing: Information Office of the State Council of People's Republic of China), english.www.gov.cn/archive/white_paper/2016/12/28/content_28147 5527159496.htm.

Notably, the Chinese military, namely the People's Liberation Army (PLA), has also given unprecedented attention to space warfare. In 2007, the PLA successfully conducted a direct-ascent anti-satellite (ASAT) test, making China the third country in the world to possess such capabilities after the US and Russia. In more recent years, the PLA has also made great efforts to develop other counterspace capabilities including co-orbital killer-satellites, lasers, jammers and other directed-energy weapons. The PLA's interests in space warfare were given a further major boost in 2015. In that year, China's Central Military Commission announced the establishment of a new Strategic Support Force (SSF) alongside the Army, Air Force, Navy and Rocket Force. The newly established SSF are entrusted with missions to undertake space, cyber and electronic warfare.

This chapter examines the PLA's views of space warfare, the development of Chinese military space capability and force structure, and emerging strategy and operational doctrines. The chapter has four sections. The first section discusses the rapidly elevated role of space in China's national military strategy. It shows that China's expanding space interests, changing perception of modern warfare and a heightened sense of insecurity in space have resulted in growing attention to the role of space in national defence planning. The second section explores the Chinese concepts of space warfare and the PLA's discussion of space strategy and operational doctrines. The third section briefly discusses the development of China's military space capability and force structure. The last section concludes with some observations.

The rising role of space in China's military strategy

China became a spacefaring country in 1970 when it successfully launched the *Dongfang Hong 1* (East is Red 1) satellite. For a long time, however, the Chinese space program had been driven more by political consideration of national prestige and economic interest in developing space-related industries and technologies than by military purposes.[6] Consequently, until the 1990s, space had been largely absent in China's national defence policy

---

6   Gregory Kulacki and Jeffrey G Lewis, *A Place for One's Mat: China's Space Program, 1956–2003* (Cambridge: American Academy of Arts and Sciences, 2009).

and military planning.[7] Indeed, China has been a longstanding advocate of the peaceful use of space and a vocal opponent to the weaponisation of outer space.

Over the last two decades or so, however, the PLA has increasingly viewed space as a new strategic domain of vital importance to national security and development, paying greater attention to the role of space in its military planning and force development. At the official level, China's 2015 Defence White Paper, entitled *China's Military Strategy*, declares that 'Outer space has become a commanding height in international strategic competition.' Judging that 'countries concerned are developing their space forces and instruments', the white paper states that as part of the PLA's force development in 'critical security domains' (重大安全领域)[8]:

> China will keep abreast of the dynamics of outer space, deal with security threats and challenges in that domain, and secure its space assets to serve its national economic and social development, and maintain outer space security.

China's 2019 Defence White Paper further assesses 'outer space security' as 'the strategic assurance of national and social development', listing protection of China's space security as one of the key tasks of the PLA. The white paper particularly highlights the importance of strengthening the capabilities of 'space situation awareness, safeguarding space assets' and 'safely enter[ing], exit[ing] and openly utiliz[ing] the outer space'.[9] Chinese military analysts are even more blunt about the important military role of outer space. A generally held view among them is that outer space has not only become a new strategic domain for national competitions but also a decisive battleground in future warfare. In 2013, the Chinese Academy of Military Science (AMS), the PLA's premier think tank, published a *Textbook for the Study of Space Operations* (空间作战学教程) for its master programs, designated for training 'high quality senior military researchers and senior staff officers' (双高人才). In the preface of the textbook, the authors noted that while peaceful use of outer space is an international consensus, competition for military space superiority has never ceased. Claiming that 'space warfare is no longer a "new fairy tale" quietly performed on the stage

---

7   For a review of the role of space in China's evolving military thinking, see Deane Cheng, 'China's Military Role in Space', *Strategic Studies Quarterly* 12, no. 1 (Spring 2012): 55–77.

8   *China's Military Strategy* (Beijing: Information Office of the State Council of People's Republic of China, 2015), english.www.gov.cn/archive/white_paper/2015/05/27/content_281475115610833.htm.

9   *China's National Defense in the New Era* (White paper) (Beijing: Information Office of the State Council of People's Republic of China, 2019), english.scio.gov.cn/2019-07/24/content_75026800.htm.

of warfare, but has become a "regular drama" frequently staged in modern warfare', the authors argued that space warfare now plays a decisive role in determining the outcomes of wars.[10] Such a view is echoed by a prominent Chinese strategist, Zhang Shibo, president of the PLA's National Defence University (NDU). Zhang argued in a 2016 book entitled *New Highland of War* that as the militarisation of outer space is accelerating in the contemporary world, gaining space superiority will have a decisive impact on both the process and outcome of modern wars.[11] He further claimed that 'space security is not only an important part of the national security, but also the pre-requisite of national security; without space security, a country will have no national security'.[12]

The elevated role of space in Chinese military thinking is a direct result of three interrelated developments. The first is the PLA's changing perception of the nature and form of modern warfare. Since the early 1990s, China has made a number of major adjustments to its military strategy in response to the Revolution in Military Affairs. It has been well documented that the PLA, based on their intensive studies of US military operations including the first Gulf War, the Kosovo Crisis and wars in Afghanistan and Iraq, has concluded that modern warfare is rapidly moving to 'informationised war'.[13] According to an authoritative PLA publication, 'informationised war' is defined as:

> war that relies on the internet-based information system, employs information-technology enabled weapons and the associated operational tactics, and is conducted mainly in the form of 'system-vs.-system warfare' [体系对抗] across land, sea, air, space, cyberspace, electromagnetic and cognitive domains.[14]

Accordingly, PLA strategists see that gaining 'information dominance' (制信息权), that is, the ability to acquire, use and control information while denying adversaries the ability to do so, is central to the wining of modern wars.

---

10   Jiang Lianju and Wang Liwen, eds, 空间作战学教程 [Textbook for the Study of Space Operations] (Beijing: Military Science Press, 2013), 1.

11   Zhang Shibo, 战争新高地 [New Highland of War] (Beijing: National Defence University Press, 2016), 13.

12   Zhang Shibo, 战争新高地 [New Highland of War], 15.

13   For a thorough review of the evolution of China's military strategy, see Taylor Fravel, *Active Defense: China's Military Strategy since 1949* (Princeton: Princeton University Press, 2020), doi.org/10.1515/9780691185590.

14   Liu Yazhou, ed., 当代世界军事与中国国防 [Contemporary World Military Affairs and China's National Defence] (Beijing: The Central Party School Press, 2016), 28.

In this context, it is not surprising that space has been accorded a vital role in PLA's thinking of fighting and winning 'informationised war'. For example, the NDU's 2017 edition of *Science of Military Strategy* (战略 学) asserted that: 'Space force is the glue of modern integrated battlefield and the glue of modern military systems'; thus, 'in the "System-vs-system" operations in future wars, without space information support, the battlefield will crumble and the warfighting system will be paralysed'.[15] Its authors estimate that the US military is currently dependent on space for 95 per cent of its intelligence, surveillance and reconnaissance, 90 per cent of communication, 100 per cent of navigation and positioning and 100 per cent of weather information, and that Russia currently relies on space assets for 70 per cent of its strategic intelligence and 80 per cent of military communication.[16]

Notably, Chinese military analysts see space as not just a critical 'force multiplier' through its provision of information support to operations in other domains, but also in itself a strategically important battlefield, from which military attacks can be launched against adversaries' space assets and strategic targets at land, air and sea. They thus assess that controlling space will be the prerequisite for gaining control in the conventional war domains of land, air and sea. Senior Colonel Zhou Bisong from NDU, a prolific PLA writer on space warfare, notes that due to the high flying speed of spacecrafts and the absence of national boundaries in outer space, space force can reach anywhere in outer space, and launch attacks on any targets on Earth, making it a force of 'global reach and global attack' (全球到达，全球作战).[17]

China's expanding space interests have been another fundamental driver behind the PLA's growing attention to space. China has wide-ranging and expanding economic, technological and geopolitical interests in outer space far beyond simple military considerations.[18] Indeed, a close watcher of China's space program, Namrata Goswami, argues that China's space ambitions have been driven more by its desire to access and explore the vast

---

15   Xiao Tianliang, ed., 战略学 [Science of Military Strategy] (Beijing: National Defence University Press, 2018), 140.

16   Xiao Tianliang, 战略学 [Science of Military Strategy], 140.

17   Zhou Bisong, 战略边疆 [Strategic Frontiers] (Beijing: Long March Press, 2015), 93.

18   Kevin Pollpeter, Eric Anderson, Jordan Wilson and Fan Yang, *China Dream, Space Dream: China's Progress in Space Technologies and Implications for the United States* (Washington, DC: U.S.-China Economic and Security Review Commission, 2015).

materials and energy resources in space than by military considerations.[19] In particular, she notes that China has ambitious plans for space-based solar power, lunar and asteroid mining and space settlement. PLA writings on space demonstrate that Chinese analysts generally perceive space as not only an important new domain of war but also a strategic frontier vital to China's future national development and survival due to the vast resources in space and technological development associated with space exploration.[20]

The expanding space interests of China have thus demanded the PLA to develop new capabilities to protect the country's growing space activities and assets and its ability to freely access and explore space. Since the early 2000s, the PLA's missions have expanded to protect not only China's national security interests but also its developmental interests. The latter includes especially China's expanding economic interests beyond its national border. In this context, it is not surprising that protecting China's space interests, which are critically important to both China's national security and development interests, has become a key mission of the PLA.

The growing importance attached to space warfare is also driven by a heightened sense of insecurity on the part of the PLA. Chinese military analysts are clearly aware of, and indeed harbour deep concerns about, the vast military space superiority possessed by the US. A PLA analyst observed warily that ever since 2001 the US has seen China as the major potential enemy in future space wars; thus, should future military conflicts occur, China will be facing grave threats of space war from the US.[21] Chinese writings often highlight the technological inferiority of China to the US in the space sector as a major vulnerability in China's space security. In this context, China's 2019 Defence White Paper stated warily that 'China's military security is confronted by risks from technology surprise and growing technological generation gap'.[22]

---

19 Namrata Goswami, 'China in Space: Ambitions and Possible Conflict', *Strategic Studies Quarterly* 12, no. 1 (Spring 2018): 74–97.
20 For example, see Zhou Bisong, 浩渺太空的竞相角逐 [Contestations in the Vast Outer Space] (Beijing: Military Science Press, 2015), especially 1–11.
21 Zhou Bisong, 战略边疆 [Strategic Frontiers], 118.
22 *China's National Defense in the New Era.*

China's emerging military space strategy and operational doctrine

Despite the growing attention paid to the important role of space in future warfare, there is no agreed definition of space warfare among PLA analysts. According to the 2011 edition of the Military Lexicon of the PLA, 'space war' (空间战), also called 'sky war' (天战) or 'outer space war' (外层空间战), refers to confrontations in outer space mainly between the military space forces. Such confrontations include offensive and defensive operations in outer space, and between outer space and air, ground and sea.[23] Some other PLA analysts, however, define space war more broadly. For them, space warfare includes not only offensive and defensive operations within space, and between space-based and Earth-based assets, but also the information support functions of space. Thus, employing space assets to provide information support for command, control, communication, computer and intelligence, surveillance and reconnaissance (C4ISR) is also considered an integral part of space warfare.[24] More recently, some Chinese analysts have included military operations in near space—a realm that is between 20 km and 100 km above sea level—as part of space warfare.[25]

The PLA hasn't formally released, perhaps has not yet developed, a space strategy or a space operation doctrine. This is evidenced by the often-heard calls made by PLA strategists for developing China's military space strategy as a matter of urgency.[26] Despite this, Chinese military writings on space could provide important insights into the emerging PLA thinking of how to conduct space warfare at both the strategic and operational levels. The following discussion is mainly based on two authoritative PLA textbooks. The first is the NDU's 2017 edition of *Science of Military Strategy* and the AMS's 2013 *Textbook for the Studies of Space Operations*.

The main focus of the 2017 edition of *Science of Military Strategy* is China's military strategy. It has, however, devoted a whole section to space warfare (termed as 'space military struggle' 太空军事斗争 in the book). In particular, the book outlines a set of strategic guiding thoughts (equivalent to the concept of military strategy in Western countries) for PLA's 'space

---

23   Jiang Lianju and Wang Liwen, 空间作战学教程 [Textbook for the Study of Space Operations], 5.
24   Zhou Bisong, 浩渺太空的竞相角逐 [Contestations in the Vast Outer Space], 144.
25   Feng Songjiang, 经略临近空间:大国战略竞争的新制高点 [Managing the Near Space: New Commanding Height of Great Power Competitions] (Beijing: Shishi Press, 2019).
26   For example, Zhang Shibo, 战争新高地 [New Highland of War], 48.

military struggles'. According to the book, China's military space strategy should include four key principles: 'combination of deterrence and real combat, limited use of space operations, comprehensive check and balance, striving for space dominance' (摄战结合，有限应对，联合制衡，争夺天权).[27]

Despite widespread external concerns about China's military space power, the principle of 'combination of deterrence and real combat' actually denotes a very cautious approach on the part of the PLA to space warfare. According to its authors, the essence of the 'combination of deterrence and real combat' is deterrence. Accordingly, deterrence should be the primary consideration in space military struggles and combat operations should play a supplementary role to support deterrence. Any use of military space power should be aimed at reinforcing the effect of deterrence. Thus, the fundamental objective of space military struggles is to prevent war through deterrence (以摄止战) in order to maintain the overall stability of the space security environment.

Such a focus on deterrence reflects a widely held view among PLA analysts that the space force is first and foremost a strategic force of deterrence. However, Chinese strategists see the impact of space operations as more controllable than that of nuclear weapons, so space force is more likely to be used in a real military confrontation than nuclear force. Given the high vulnerability of space assets and lower threshold of space operations, space deterrence is perceived as more credible and more likely to be used than nuclear deterrence.[28]

The second principle, 'Limited use of space operations', means that at the strategic level, any space operations should be undertaken in a limited way, leaving room for de-escalation (留有余地). Those limited space operations should primarily serve two goals: ensuring effective deterrence against an adversary stronger than oneself and acquiring space superiority over an opponent weaker than oneself. The limited use of space operation, thus, will allow a nation to exercise strategic initiatives to control the course and outcome of space war. Such a principle is in accord with the above-mentioned overriding focus on deterrence.

---

27 Xiao Tianliang, ed., 战略学 [Science of Military Strategy], 144.
28 Zhang Shibo, 战争新高地 [New Highland of War], 19.

'Comprehensive check and balance' requires a holistic approach to space military struggles. It is based on the recognition that space security is an issue influenced by myriad factors, including, but not limited to, the interdependence of nation-states in relation to their space interests and security, and the mutual influences between space assets and between space systems and ground facilities. It is thus important to take a comprehensive approach, using various means to deter and contain the external threats to space security. This could involve both military and non-military means, and operations not only in outer space but in other domains as well.

'Striving for space dominance' stipulates that the overriding goal of space military struggle is to acquire 'space dominance' (制天权). For the PLA, 'space dominance' means a country's unfettered access to and use of space while denying an adversary's access.[29] In this context, in the inevitable future space confrontations, it is important to ensure the stability of one's space system through continuous, undisrupted space logistics, support and combat operations.

While the above strategic guidelines demonstrate a cautious approach to space warfare, at the operational level, PLA writings show a strong preference for taking initiative and striking first. The AMS's 2013 *Textbook for the Study of Space Operations* provides extensive discussions on the conduct of space operations. Among others, it describes the guiding principles of China's space operations as 'active defence, integration of all dimensions, focused space dominance' (积极防御，全维一体，重点制天).[30]

'Active defence' is the defining principle of China's general military strategy. Officially speaking, it denotes an overall strategic defensive posture for the Chinese military. As China's 2015 Defence White Paper explains, a key aspect of 'Active Defence' is that 'We [China] will not attack unless we are attacked, but we will surely counterattack if attacked.' The defensive nature of China's military strategy, however, doesn't prevent China from attacking first at the campaign and tactical level, as 'active defence' also means the 'unity of strategic defence and operational and tactical offence'.[31] The latter, even if offensive in nature, is perceived as justified as long as its aim is to

---

29   Guo Rongwei, 九天揽月: 中国太空战略发展研究 [Catching the Moon in the Sky: A Study of the Development of China's Space Strategy] (Beijing: National Defence University Press, 2014), 245.
30   Jiang Lianju and Wang Liwen, 空间作战学教程 [Textbook for the Study of Space Operations], 40.
31   *China's Military Strategy.*

achieve strategic defence. Thus, while the PLA will not make the first move to start a war, it intends to seize initiatives at the operational level when war occurs.

In this context, the *Textbook of the Study of Space Operations* stated explicitly that the key to implementing the principle of 'active defence' is to establish the view of 'striking first in space warfare' once an overall war occurs.[32] Notably, it argued that as informational support of space is now an integral part of modern warfare, when an adversary starts to use space assets for surveillance, reconnaissance and intelligence collections in the context of an imminent war, it can be deemed that the adversary 'makes the first shot' in the war. It is justified for China to take the initiative to launch operational attacks first on the adversary's space system even in this war preparation stage. It is also argued that such attacks should be done through 'sudden and quick' operations to achieve the effect of 'space blitzkrieg'.[33]

'Integration of all dimensions' requires that all space operations should be undertaken in a joint manner involving the operation in all domains including land, sea, air, space, cyberspace and even legal and psychological warfare. Thus, space operations are expected to be conducted in coordination with operations in other realms so as to achieve the desired strategic outcomes. Moreover, 'Integration of all dimensions' also requires collaborations between the military and civilian sectors, through the integrated use of military, civilian and commercial space assets for space operations, and sees the civilian and commercial space system forces as important elements of China's space warfare capability.

Like the guiding principles of the military space strategy as discussed above, 'Focused space dominance' also highlights the importance of gaining control of space. The principle, however, provides more specific requirements and guidance for space warfare at the operational level. While the essential goal of space operations is to gain 'space dominance', the key to this principle is 'focused' space dominance or space dominance with a specific focus (有重点的制天). Accordingly, space operations should identify and attack 'key targets' of the adversary's military space system through tactics of disruption, denial, damage and destruction, so as to remove the adversary's space

---

32 Jiang Lianju and Wang Liwen, 空间作战学教程 [Textbook for the Study of Space Operations], 40–41.
33 Jiang Lianju and Wang Liwen, 空间作战学教程 [Textbook for the Study of Space Operations], 45–46, 51–52.

capabilities.[34] In this context, gaining 'space dominance' doesn't require, and indeed should not aim to have, complete control of space. Rather it refers to a 'limited and targeted' control of space. This involves, for example, deploying an elite space force, attacking the most important targets with carefully chosen timing. Accordingly, the objective of space operations is not the total destruction of the adversary's military space system, but to disable or disrupt the enemy's information chain, achieving a 'nerve-cutting' effect at critical points of time, to deny the enemy's access to the relevant space systems for limited time.[35]

The emphasis on 'focused space dominance' also leads to a preference for 'soft kill' methods over 'hard kill' in space operations. The former refers to the use of non-kinetic counterspace means such as low-energy lasers, electronic and magnetic weapons, and cyber operations to attack an enemy's space system. Such attacks aim to disable, disrupt and degrade the enemy's space capabilities temporarily, but not destroy them physically. The advantage of 'soft kill' methods is that they are hard to detect, difficult to trace back to the source and can effectively conceal one's operational intention, thus making them an important means of gaining space dominance. In contrast, 'hard kill' attacks, which seek to destroy an enemy's space assets, are often riskier and could lead to unintended consequences such as unwanted escalations of war. It could also cause collateral damage to the entire space system, including the attacker's own space assets, due to the debris caused by the physical attacks.[36]

China's space military capability and force structure

A major development in China's space warfare capability and force structure is the establishment of the Strategic Support Force (SSF) on 31 December 2015, as part of the sweeping military reforms initiated by President Xi Jinping in his effort to build the PLA into a 'world class' military.[37] The SSF

---

34 Jiang Lianju and Wang Liwen, 空间作战学教程 [Textbook for the Study of Space Operations], 41.

35 Jiang Lianju and Wang Liwen, 空间作战学教程 [Textbook for the Study of Space Operations], 45.

36 Jiang Lianju and Wang Liwen, 空间作战学教程 [Textbook for the Study of Space Operations], 47–48.

37 Jian Zhang, 'Toward a "World Class" Military: Reforming the PLA under Xi Jinping', in *China Story Yearbook 2019: Power*, ed. Jane Golley, Linda Jaivin, Paul J Farrelly, and Sharon Strange (Canberra: ANU Press, 2019), 218–230, doi.org/10.22459/CSY.2019.08.

brings together most of the PLA's space, cyber, electronic, and psychological warfare capabilities that were previously dispersed in various services and departments into a single integrated independent force.[38] While the SSF is not formally listed as an independent service like the Army, the Air Force, the Navy and the newly established Rocket Force, it enjoys a de facto service-level status and is under the direct control of China's Central Military Commission.

According to Chinese military analysts, unlike the traditional military force whose main capability is to destroy and kill, the SSF's main power comes from its ability to disrupt and degrade the functions of the adversary's C4ISR system and decision-making process. Accordingly, SSF's missions are broad, including space operations, surveillance and intelligence gathering, cyberwarfare, electronic warfare and, according to some analysts, even psychological warfare.[39]

Space warfare occupies a particularly important position in SSF's activities. At the operational level, the SSF comprises two main departments: the Space Systems Department which is responsible for space operations, and the Network Systems Department which leads cyber operations, electronic warfare and potentially psychological warfare.[40] The Space System Department, a deputy theatre command level body, commands and controls the PLA's military space force, which includes units and bases responsible for space launch and support, space telemetry, tracking and control, navigation, space information support, space attack and space defence.[41]

Despite the PLA's efforts to develop an integrated separate space force under SSF, the current reorganisation is perhaps at best a 'work in progress'. It appears that a significant part of China's space warfare capabilities is still controlled by the PLA Rocket Force (PLARF) and PLA Air Force (PLAAF) separately. For example, it is currently not clear whether SSF or PLARF and PLAAF are responsible for the development of China's anti-satellite capabilities and ballistic missile defence (BMD) systems, or what their respective roles are in those areas. The PLARF, as China's strategic nuclear

---

38   John Costello and Joe McReynolds, *China's Strategic Support Force: A Force for a New Era*, China Strategic Perspectives No. 13 (Washington, DC: National Defence University, October 2018), 22.

39   Costello and McReynolds, *China's Strategic Support Force*, 22.

40   Costello and McReynolds, *China's Strategic Support Force*, 20–28. It should be noted that due to the integrated nature of information operations, there might be overlapping activities and functions between the two departments.

41   Costello and McReynolds, *China's Strategic Support Force*, 20–28.

force, commands and controls both the nuclear and conventional missile capabilities and has long played an important role in the development of ASAT and BMD capabilities. The PLAAF has also demonstrated a growing interest in space warfare. In 2004, it formally announced the goal of transforming itself from a conventional territorial-based defence-oriented air force into a strategic force with integrated air and space offensive and defensive capabilities.[42] This requires, as advocated by some PLA analysts, that the PLAAF develop space and counterspace capabilities.[43] Indeed, in November 2018, Lieutenant General Xu Anxiang, the deputy commander of PLAAF, flagged the goal of establishing such a strategic air force by 2020.[44]

In addition to force structure changes, the PLA has also made significant progress in its military space capabilities over the past two decades. Compared with the US and Russia, China is a relative latecomer in military space. Since 2000, however, the country has achieved tremendous progress in building up one of the world's most extensive space warfighting systems. Consequently, as of mid-2016, China had already developed the world's second-largest satellite system after the US, comprising a total of 181 satellites from an initial handful of satellites.[45] Less than two years later, the number of Chinese satellites in space jumped to 299 as of March 2019.[46]

Accordingly, the PLA has achieved enormous progress in the development of space-based reconnaissance, navigation and communications capabilities. It was found that with an almost nil space capacity for living surveillance at the beginning of the 2000s, the PLA today possessed significant space-based reconnaissance capabilities, building on one of the world's largest and rapidly growing constellations of remote sensing satellites of various sorts.[47] China has also successfully developed its indigenous global navigation satellites system (GNSS), the Beidou system, as a rival to the US-operated global positioning system (GPS). The Beidou system consists of 24 navigation

42   Michael Chase and Cristina Garafola, 'China's Search for a "Strategic Air Force"', *Journal of Strategic Studies* 39, no. 1 (2016): 4–28, doi.org/10.1080/01402390.2015.1068165.

43   Chase and Garafola, 'China's Search'.

44   SpaceWatch Asia Pacific, 'Future Space Wars: China's PLAAF Intends to Extend Its Reach into Space', *SpaceWatch Asia Pacific*, November 2018, spacewatch.global/2018/11/future-space-wars-chinas-plaaf-intends-to-extend-its-reach-into-space/.

45   Kevin L Pollpeter, Michael S Chase and Eric Heginbotham, *The Creation of the PLA Strategic Support Force and Its Implications for Chinese Military Space Operations* (Santa Monica: Rand Corporation, 2017), 8, doi.org/10.7249/RR2058.

46   China Power, 'How Is China Advancing Its Space Launch Capabilities?'.

47   Eric Hagt and Matthew Durnin, 'Space, China's Tactical Frontier', *Journal of Strategic Studies* 34, no. 5 (2011): 733–761, doi.org/10.1080/01402390.2011.610660.

satellites and has been fully operational to provide complete global service since the end of 2019.[48] This indigenous GNSS substantially increased both the security and capability of China's military space program. China also has an expanding fleet of communication satellites for both civilian and military use.

What is of most concern to outside observers, however, is China's rapidly developing counterspace capabilities. While China's anti-satellite test in 2007 drew worldwide attention, it is just one part of China's wide-ranging counterspace military program. It is noted that the PLA has been developing various kinetic and non-kinetic counterspace capabilities including missiles, co-orbital satellite killers, directed-energy weapons, jammers, and electromagnetic and cyber weapons programs.[49] Given the noted US reliance on satellites for its military operation, it is not surprising that the development of anti-satellite capabilities has been perceived as the top priority area for PLA's counterspace program.[50]

Conclusion

Space warfare has clearly assumed an important role in China's military planning and force development. The PLA's changing perception of modern warfare has led to the recognition of space as the critical new domain of 'informationised war'. China's wide-ranging and expanding interests in the use and explorations of outer space and its deeply harboured concerns about space security have further contributed to the elevated role of space in national defence policy and military development. Despite the increasingly important role accorded to space in future warfare, China's emerging military strategy and operational doctrine have demonstrated a cautious approach to space warfare, a focus on deterrence and a preference for limited use of space operations. Having said that, the extraordinary progress made by China in the development of its space and counterspace capabilities over the last two decades has made it one of the most powerful players in future space warfare.

---

48   Stephen Clark, 'China Completes Core of Beidou Global Satellite Navigation System', *Spaceflight Now*, 16 December 2019, spaceflightnow.com/2019/12/16/china-completes-core-of-beidou-global-satellite-navigation-system/.

49   Pollpeter, Chase and Heginbotham, *The Creation of the PLA Strategic Support Force*, 9.

50   Zhou Bisong, 浩渺太空的竞相角逐 [Contestations in the Vast Outer Space], 311.

References

Bowe, Alexander. *China's Pursuit of Space Power Status and Implications for the United States*. Staff Research report. U.S.–China Economic and Security Review Commission, 11 April 2019).

Chase, Michael, and Cristina Garafola. 'China's Search for a "Strategic Air Force"'. *Journal of Strategic Studies* 39, no. 1 (2016): 4–28. doi.org/10.1080/01402390. 2015.1068165.

Cheng, Deane. 'China's Military Role in Space'. *Strategic Studies Quarterly* 12, no. 1 (Spring 2012): 55–77.

China Power. 'How Is China Advancing Its Space Launch Capabilities?' *China Power*, 5 November 2019. chinapower.csis.org/china-space-launch/.

China's Military Strategy. Beijing: Information Office of the State Council of People's Republic of China, 2015. english.www.gov.cn/archive/white_paper/2015/05/ 27/content_281475115610833.htm.

China's National Defense in the New Era. White paper. Beijing: Information Office of the State Council of People's Republic of China, 2019. english.scio.gov.cn/ 2019-07/24/content_75026800.htm.

China's Space Activities in 2016. White paper. Beijing: Information Office of the State Council of People's Republic of China, 2016. english.www.gov.cn/archive/ white_paper/2016/12/28/content_281475527159496.htm.

Clark, Stephen. 'China Completes Core of Beidou Global Satellite Navigation System'. *Spaceflight Now*, 16 December 2019. spaceflightnow.com/2019/12/16/ china-completes-core-of-beidou-global-satellite-navigation-system/.

Costello, John, and Joe McReynolds. *China's Strategic Support Force: A Force for a New Era*. China Strategic Perspectives No. 13. Washington, DC: National Defence University, October 2018.

Feng Songjiang. 经略临近空间:大国战略竞争的新制高点 [Managing the Near Space: New Commanding Height of Great Power Competitions]. Beijing: Shishi Press, 2019.

Fravel, Taylor. *Active Defense: China's Military Strategy since 1949*. Princeton: Princeton University Press, 2020. doi.org/10.1515/9780691185590.

Goswami, Namrata. 'China in Space: Ambitions and Possible Conflict'. *Strategic Studies Quarterly* 12, no. 1 (Spring 2018): 74–97.

Guo Rongwei. 九天揽月: 中国太空战略发展研究 [Catching the Moon in the Sky: A Study of the Development of China's Space Strategy]. Beijing: National Defence University Press, 2014.

Hagt, Eric, and Matthew Durnin. 'Space, China's Tactical Frontier'. *Journal of Strategic Studies* 34, no. 5 (2011): 733–761. doi.org/10.1080/01402390.2011.610660.

Jiang Lianju and Wang Liwen, eds. 空间作战学教程 [Textbook for the Study of Space Operations]. Beijing: Military Science Press, 2013.

Jones, Andrew. 'China to Continue World-Leading Lauch Rate in 2020'. *Space News*, 3 December 2019. spacenews.com/china-to-continue-world-leading-launch-rate-in-2020/.

Kulacki, Gregory, and Jeffrey G Lewis. *A Place for One's Mat: China's Space Program, 1956–2003*. Cambridge: American Academy of Arts and Sciences, 2009.

Li Yazhou, ed. 当代世界军事与中国国防 [Contemporary World Military Affairs and China's National Defence]. Beijing: The Central Party School Press, 2016.

Pollpeter, Kevin, Eric Anderson, Jordan Wilson, and Fan Yang. *China Dream, Space Dream: China's Progress in Space Technologies and Implications for the United States*. Washington, DC: U.S.-China Economic and Security Review Commission, 2015.

Pollpeter, Kevin L, Michael S Chase, and Eric Heginbotham. *The Creation of the PLA Strategic Support Force and Its Implications for Chinese Military Space Operations*. Santa Monica: Rand Corporation, 2017. doi.org/10.7249/RR2058.

SpaceWatch Asia Pacific. 'Future Space Wars: China's PLAAF Intends to Extend Its Reach into Space'. *SpaceWatch Asia Pacific*, November 2018. spacewatch.global/2018/11/future-space-wars-chinas-plaaf-intends-to-extend-its-reach-into-space/.

Xiao Tianliang, ed. 战略学 [Science of Military Strategy]. Beijing: National Defence University Press, 2018.

Zhang, Jian. 'Toward a "World Class" Military: Reforming the PLA under Xi Jinping'. In *China Story Yearbook 2019: Power,* edited by Jane Golley, Linda Jaivin, Paul J Farrelly, and Sharon Strange, 218–231. Canberra: ANU Press, 2019. doi.org/10.22459/CSY.2019.08.

Zhang Shibo. 战争新高地 [New Highland of War]. Beijing: National Defence University Press, 2016.

Zhou Bisong. 浩渺太空的竞相角逐 [Contestations in the Vast Outer Space]. Beijing: Military Science Press, 2015.

Zhou Bisong. 战略边疆 [Strategic Frontiers]. Beijing: Long March Press, 2015.

7

Quantum Technologies: An Introduction and a Vision of their Impact on War

Marcus Doherty

Introduction[1]

Quantum technologies exploit the fundamental laws of nature to reach the ultimate limits of sensing, imaging, communications and computing, and thus promise otherwise impossible capabilities.[2] They are no longer scientific speculation; substantial public and private investments around the world are accelerating their development and application.[3] This acceleration will likely see quantum technologies transform defence, science and industry over the next 20 years, particularly when combined with other emerging technologies, such as nanotechnology, biotechnology, space technology, artificial intelligence (AI) and robotics. Yet precisely when and how the quantum technologies will transform defence is not clear.

---

1   Editors' note: as mentioned in the Introduction, this chapter was written in 2020, and global events since then may have overtaken some aspects. As the pace of technological and cultural change has continued to accelerate, the editors have opted to present this text as drafted to minimise further delays in publication.

2   Jason Palmer, 'Here, There and Everywhere; Technology Quarterly', *The Economist*, 11 March 2017.

3   Elizabeth Gibney, 'Quantum Gold Rush: The Private Funding Pouring into Quantum Start-Ups', *Nature* 574, no. 7776 (2019): 22, doi.org/10.1038/d41586-019-02935-4.

Quantum technologies are diverse, complex and demand new ways of thinking about the employment and exploitation of technology.[4] Their true capabilities, limitations and most disruptive applications are still being discovered. This ambiguity and complexity present both strategic risks and opportunities to defence forces. They are in an intensifying global competition to understand, co-develop and exploit quantum technologies in military operations.

To help defence professionals meet this demanding challenge, this chapter provides:

- an introduction to quantum technologies: their definition, types, characteristics, capabilities, limitations, defence applications and projected development timelines
- a vision and assessment of their employment and impact in warfare over the next 20 years.

Note that this chapter does not contain in-depth technical details of quantum technologies, which are best found elsewhere.

Quantum technologies

Quantum technology is a term that encompasses a diverse suite of technologies. These technologies are at different levels of readiness and have different development timelines. Their defence applications are rapidly evolving and expanding, with many yet to be discovered. The aim of this section is to introduce the key information required to navigate quantum technologies and to begin identifying and assessing their defence applications.

---

4   Department of Defence, *Army Quantum Technology Roadmap* (Canberra: Australian Government Publishing Service, 2021).

Definition

A quantum technology is one whose functionality derives from engineering the states of quantum systems.[5] This distinguishes quantum technologies from the various 20th-century technologies (e.g. lasers, magnetic resonance imaging, semiconductor electronics) that employ quantum phenomena (e.g. coherence, quantised energy, tunnelling), but do not directly initialise, manipulate and measure the states of individual quantum systems.

A quantum system is a system of elementary particles (e.g. electrons, photons and nuclei) whose behaviour is governed by the laws of quantum physics.[6] Measurements of a quantum system have random values whose probabilities are determined by the system's state at the time of measurement. Following measurement, a quantum system is projected into a state that matches the measurement mechanism and value. A quantum system's state at a given moment can be described as a superposition of the states associated with a measurement mechanism: the simultaneous occupation of multiple states with definite relative amplitudes and phases. Some states exhibit entanglement of two or more subsystems (i.e. subgroups of particles) of the quantum system. Entanglement produces statistical correlations in the values of measurements of the individual subsystems. Interactions between a quantum system and its environment can randomise its state. This process is called decoherence and it ultimately limits how precisely a quantum system's state can be engineered.

The fundamental building block of quantum technology is the qubit: the simplest quantum system, which has two states: $|0\rangle$ and $|1\rangle$.[7] The qubit is a useful abstract concept that allows us to understand how different quantum technologies work and compare. In practice, different systems of particles or different variables of similar systems play the role of the qubit in different technologies.

5   Jonathan P Dowling, and Gerard J Milburn, 'Quantum Technology: The Second Quantum Revolution', *Philosophical Transactions of the Royal Society A* 361, no. 1809 (2003): 1655, doi.org/10.1098/rsta.2003.1227.

6   For a popular introduction to quantum physics and qubits, see Leonard Susskind and Art Friedman, *Quantum Mechanics: The Theoretical Minimum* (New York: Basic Books, 2015).

7   For a more advanced understanding of qubits and quantum technologies, see Michael A Nielsen and Isaac L Chuang, *Quantum Computation and Quantum Information*, 10th anniversary edition (Cambridge: Cambridge University Press, 2010).

Characteristics

Quantum technologies function by using normal 'classical' devices (e.g. lasers, microwave electronics and photodetectors) to initialise (e.g. by a laser pulse), manipulate (e.g. by a microwave pulse) and measure (e.g. by detecting emitted photons) the state of their qubit(s).[8] A normal 'classical' computer is used to program and control these devices and record the measurement data. Thus, quantum technologies are operated via a familiar classical computing interface. Despite the qubits being a relatively small component of the quantum technology, their different physical behaviour is what delivers an advantage over classical technologies.

The performance of a quantum technology is determined by both its qubits and its classical control system and methods.[9] Performance is improved by engineering higher-quality qubit systems, classical control hardware and methods, and shielding the qubits from the environment. The first three are pushing the limits of material growth, microfabrication, electrical, optical and mechanical engineering, and optimal control design. Environmental shielding requirements depend on the type of technology and qubit system. Some require cryogenics to achieve ultra-low temperatures and/or vacuum systems to achieve ultra-high mechanical isolation. Others don't require either and can operate in ambient and extreme conditions. Nevertheless, high-quality materials, fabrication and device engineering are key to high-performance quantum technology.

Types and timelines

Thus far, quantum technologies can be broadly categorised into three main types:

a. quantum sensing and imaging—new limits in precision and stability

b. quantum communications and cryptography—networking quantum devices and physically assured security

c. quantum computing—a leap in computational power for certain tasks.

---

8   This section was first published in Marcus W Doherty, 'Quantum Technology: An Introduction', *Land Power Forum, Australian Army Research Centre,* 27 May 2020, researchcentre.army.gov.au/library/land-power-forum/quantum-technology-introduction.

9   Doherty, 'Quantum Technology: An Introduction'.

The current development states and projected future development timelines of technologies within each type are summarised in Table 7.1.[10] There, the technologies are broadly organised into three readiness categories: in production/advanced stages of industry research and development, in intermediate stages of industry/advanced stages of academic research and in early stages of research, with respective estimates of when they will be ready for defence applications of 0–5 years, 5–10 years and 10-plus years.

Table 7.1: Current state and projected development timelines of different quantum technologies.

Current development state	In production/ advanced stages of industry R&D	Intermediate industry R&D/ advanced stages of academic research	Early stages of industry R&D/intermediate stages of academic research
Estimated time to defence application	**<5 years**	**5–10 years**	**>10 years**
Sensing & imaging	· Quantum accelerometers, magnetometers, gyroscopes and clocks · Quantum microscopes	· Quantum spectrometers and detectors · Chip-scale bio/chemical analysers · Quantum-enhanced MRI	· Wearable magnetoencephalography · Quantum nanosensors for biomedicine

10   As synthesised by the author from: Frank Arute, et al., 'Quantum Supremacy Using a Programmable Superconducting Processor', *Nature* 574 (2019): 505; Elena Boto, et al., 'A New Generation of Magnetoencephalography: Room Temperature Measurements Using Optically-Pumped Magnetometers', *NeuroImage* 149 (2017): 404, doi.org/10.1016/j.neuroimage.2017.01.034; Edwin Cartlidge, 'Quantum Sensors: A Revolution in the Offing?' *Optics & Photonics News* (September 2019): 24–31, doi.org/10.1364/OPN.30.9.000024; CL Degen, F Reinhard, and P Cappellaro, 'Quantum Sensing', *Reviews of Modern Physics* 89 (2017): 035002, doi.org/10.1103/RevModPhys.89.035002; Marcus W Doherty, 'Quantum Microscopy', *AOS News* 30 (2016): 2; Marcus W Doherty, Andre Luiten, David Simpson, and Liam Hall, 'Quantum Technology: Sensing and Imaging', *Land Power Forum, Australian Army Research Centre,* 2 September 2020, researchcentre.army.gov.au/library/land-power-forum/quantum-technology-sensing-and-imaging; Doherty, 'Quantum Technology: An Introduction'; Dowling and Milburn, 'Quantum Technology: The Second Quantum Revolution'; Susskind and Friedman, *Quantum Mechanics*; Jay Gambetta, 'IBM's Roadmap For Scaling Quantum Technology', *IBM Research Blog*, 15 September 2020, www.ibm.com/blogs/research/2020/09/ibm-quantum-roadmap/; Julian Kelly, 'A Preview of Bristlecone, Google's New Quantum Processor', *Google AI Blog*, 5 March 2018, ai.googleblog.com/2018/03/a-previewof-bristlecone-googles-new.html [site discontinued]; Nielsen and Chuang, *Quantum Computation and Quantum Information*; Palmer, 'Here, There and Everywhere'; John Preskill, 'Quantum Computing in the NISQ Era and beyond', *Quantum* 2 (2018): 79, doi.org/10.22331/q-2018-08-06-79; Stephanie Wehner, David Elkouss, and Ronald Hanson, 'Quantum Internet: A Vision for the Road Ahead', *Science* 362, no. 6412 (2018): eaam9288, doi.org/10.1126/science.aam9288.

Communications & cryptography	• Simple short-range QKD networks	• Quantum repeaters • Quantum ports • Complex long-range QKD networks • Synchronisation of clocks	• Quantum memories • Networks of quantum sensors and computers
Computing & simulation	• Mainframe NISQ computers	• Distributed and edge NISQ computers integrated in classical networks • EC mainframe computers	• Large-scale EC mainframe computers capable of cryptography • Distributed EC computers in quantum networks

Notes: EC = error corrected; MRI = magnetic resonance imaging; NISQ = noisy intermediate-scale quantum; QKD = quantum key distribution; R&D = research and development.

Source: Author's synthesis of multiple sources, see footnote 10.

The following subsections describe the operating principles, capabilities, limitations and preliminary defence applications of each technology type, as well as the key classical technologies that enable them to function and possible countermeasures that defeat them.

Quantum sensing and imaging

Quantum sensors measure time, dynamics (i.e. forces, acceleration and rotation) and fields (i.e. gravitational, electromagnetic and mechanical) with unprecedented precision and stability.[11] Imaging is an extension of quantum sensing where quantum sensors are combined with an imaging apparatus (e.g. a probe that scans the position of the sensor, an array of sensors or a beam of electromagnetic waves prepared in a quantum state) to perform high-resolution microscopy or macroscopy (e.g. radar) with unprecedented sensitivity.

Quantum sensors exploit quantum superposition to apply interferometry techniques to detect small changes in a qubit's state by the passage of time, dynamics or interactions with fields. Quantum entanglement between

---

11   For an introduction to quantum sensing, see Cartlidge, 'Quantum Sensors: A Revolution in the Offing?' A comprehensive technical review of quantum sensing is Degen, Reinhard, and Cappellaro, 'Quantum Sensing'.

multiple qubits may be exploited to further enhance precision. Stability is achieved through the qubits having fixed and universal susceptibilities (e.g. electron gyromagnetic ratio and atomic mass):

> The limitations of quantum sensors are primarily dynamic range, speed and environmental exposure. Although highly sensitive, quantum sensors are often limited to a small range of measurand values and will saturate if there are large variations ... Quantum sensors can take longer to perform measurements than other technologies, and so cannot provide the same update rate. The environment can bring unwanted noise that deteriorates the qubit properties and so the qubits need to be shielded from the elements of the environment other than the measurand. This is the classic packaging problem experienced by all sensors. The first two limitations can be ameliorated by integrating quantum sensors with other sensors that are faster and have larger dynamic ranges. The third limitation is a difficult problem, but one where there is an established wealth of knowledge to draw upon, and thus reason to be optimistic.[12]

The most promising defence applications[13] are:

a. *Enhanced positioning, navigation and timing.* Quantum accelerometers, magnetometers, gyroscopes and clocks promise the long-term stability and precision required for accurate inertial navigation in the absence of global positioning systems (GPS). Furthermore, advanced atomic clocks promise precision timing that is key to accurate positioning as well as reducing errors and increasing the speed of digital communications.

b. *Enhanced situational awareness and targeting.* Quantum gravimeters and magnetometers promise new capabilities in geospatial mapping and anomaly detection. For example, the detection of subterranean structures or objects. Quantum microwave spectrometers promise high sensitivity and unbroken detection bands for monitoring the electromagnetic spectrum. These could be used to enhance the range and precision of existing radar systems or, if networked, used to enhance the detection and locating of electromagnetic emitters.

c. *Enhanced medical and environmental analysis.* Quantum nanosensors offer unprecedented means to perform biological and chemical sensing and imaging for medical diagnosis and treatment and environmental

---

12 Doherty et al., 'Quantum Technology: Sensing and Imaging'.
13 This section expands upon discussion in Doherty et al., 'Quantum Technology: Sensing and Imaging'.

testing and monitoring. The technologies range from sensitive nanoparticles that are injected into a specimen and tracked/imaged, to field-deployable lab-on-a-chip biochemical analysers, to add-ons that enhance the resolution of existing medical imaging techniques, like magnetic resonance imaging (MRI).

d. *Enhanced human–machine interfacing.* Quantum magnetometers may enable wearable, high-resolution magnetoencephalography for real-time brain activity imaging: a key ingredient to practical, direct cognitive communication with machines.

e. *Enhanced defence science and industry.* Quantum microscopes, spectrometers and nanosensors promise to drive innovation in materials science and nano-, bio- and medical technology. For example, quantum microscopes are a critical characterisation tool for the development of low-energy, two-dimensional (e.g. graphene) electronics and will potentially accelerate drug design by providing a more direct and functional means to image the chemical structure of proteins and other complex molecules that are key to microbiological processes. Additionally, chip-scale nuclear magnetic resonance spectrometers promise the democratisation of precision chemical analysis, whose restricted availability is a profound constraint in current biological, chemical and material research and development.

Quantum communications and cryptography

Quantum communications can be used to network quantum sensors to correlate and enhance sensitivity over large areas (e.g. synchronise clocks within a communication network) and networking quantum computers to efficiently exchange data and amass computational power.[14] Quantum communications may also be used to securely access remote quantum computers or securely transmit data between classical devices (e.g. distribute encryption keys).[15] The latter is known as quantum key distribution (QKD) and is a means to enhance the security of conventional communications by enabling the secure distribution of one-time private encryption keys between remote nodes of a conventional encrypted network.

---

14 A useful overview of the current state and future directions in quantum communications is in Wehner, Elkouss, and Hanson, 'Quantum Internet'.

15 Wehner, Elkouss, and Hanson, 'Quantum Internet'.

Predominately, quantum communication is implemented by sending superimposed or entangled qubits between devices.[16] The qubits are fundamental particles of light (i.e. photons) and thus quantum communications can be understood as the ultimate limit of optical communications. As per normal optical communications, quantum communication can be performed between remote nodes using terrestrial optical fibre links, direct free-space (i.e. line-of-sight) links or satellite-mediated links between ground stations. Quantum repeaters are required to overcome losses in fibre networks, to synchronise communications in complicated networks and to enable over-the-horizon communication via satellite links. To fulfil this role, quantum repeaters must be able to collect (without measurement), store (for sufficient time) and regenerate qubits with high probability and fidelity. Networking quantum sensors and computers requires an additional technology: quantum ports that interface the photons acting as the communication qubits with the physical qubits of the sensor/ computer.

Alternatively, quantum communication can be achieved by the transportation and exchange of long-term quantum memories between parties, analogous to exchanging encrypted hard drives or documents.

Security is physically assured in quantum communications by the projective nature of quantum measurements, which means that it is not physically possible to copy the qubit-encoded information without modifying it.[17] Thus, security is assured in the sense that interception and interference by eavesdroppers can be detected with a known statistical confidence. The confidence level depends on the number of transmitted qubits sacrificed to distil the encryption key and amplify privacy and the noise present in the communication channel.

To date, there has been a variety of demonstrations of QKD over optical fibre networks, direct free-space links and satellite-mediated links. However, these have not included quantum repeaters and so have been limited in their range (without the use of 'trusted' intermediary nodes) and limited in the complexity of their network traffic without the ability to use repeaters to combine scheduling with routing. The creation of quantum repeaters, ports and long-lived memories is technically challenging and there is still

---

16   For a technical understanding, see Nielsen and Chuang, *Quantum Computation and Quantum Information*, Section 12.6.2, 587.

17   Nielsen and Chuang, *Quantum Computation and Quantum Information*.

some time before these will be sufficiently mature to enhance quantum communication networks. Putting the difficulty of these technologies aside, the rate of quantum communication (e.g. key generation rate in QKD or remote entanglement generation rate in networking) is currently relatively slow and presents another significant technical challenge to improve. Thus, widespread QKD or quantum networking is unlikely in the near future. Rather, quantum communications are likely to be restricted to a few high-priority links in the near future.

Defence applications of quantum communications include:

a. *Enhanced security of communication networks via QKD.* QKD offers a means to securely distribute one-time private encryption keys and so enhance the security of the encrypted communication that currently relies on public key encryption methods. This is motivated by the fact that public key encryption relies on the difficulty of solving certain mathematical problems (e.g. prime factoring) rather than a physical effect (like QKD) to maintain security. Thus, if an efficient way of solving these mathematical problems is developed (e.g. a quantum computer), the security of public key encryption is dramatically reduced—a weakness not shared by private key encryption. The trouble is that private keys should only be used once, which means that keys must be securely distributed at a sufficiently high rate to service the traffic of a network. A QKD network operating in parallel with a conventional communication network addresses this key distribution problem.

b. *Enhanced network synchronisation.* Distributing entanglement can be used as a physical means of precisely synchronising distant clocks to improve the speed and accuracy of digital communications as well as the correlation of distant classical sensors to more precisely detect and/or locate anomalies.

c. *Enhanced sensitivity and large-area quantum sensing.* Entangling multiple quantum sensors at a single location improves their combined sensitivity beyond that offered classically. Entangling distant quantum sensors enhances their ability to simultaneously correlate their measurements and so more precisely detect and/or locate anomalies.

d. *Enhanced security and performance of quantum computing.* Quantum communications can enhance the efficiency of data exchange between quantum computers, amass their computational power and enable secure access to them. Networking quantum computers is expected to enable a similar expansion in performance and applications as the internet has enabled for classical computers.

Quantum computing and simulation

Quantum computers dramatically speed up the solution of particular computational problems.[18] While the full range of such problems is still being discovered, established examples are related to signal/image processing, machine learning, optimisation, simulation, searching and factoring. They achieve this by exploiting quantum superposition and entanglement to represent and manipulate information in a fundamentally more dense and efficient way than classical computers. Thus, quantum computers require fewer physical resources and operations to solve the same problem as a classical computer.

However, it is important to note that quantum computers have a slower 'clock rate' than classical computers, which means that they won't speed up calculations that are already fast on classical computers. Rather, a better picture is that they will take seconds to perform a calculation that would take days, years or even longer on a classical computer, thereby making certain calculations practical for the first time. Another way of thinking about the advantages of quantum computing is that, for a given calculation time, quantum computers deliver a more accurate or comprehensive solution than classical computers, which is relevant to situations where accuracy is more important than speed.

The key metrics of a quantum computer are:

a. number of qubits—the amount of information that can be encoded and processed by the computer
b. circuit depth—the complexity of the algorithms that can be performed by the computer
c. speed—the time it takes to perform a computation
d. fidelity—the precision of the computation (how unlikely errors are).

---

18   For an introduction to quantum computing, see this overview: Vlatko Vedral, 'Instant Expert 33: Quantum Information', *New Scientist*, multiple dates, www.newscientist.com/round-up/quantum-information/. For a technical understanding, see Nielsen and Chuang, *Quantum Computation and Quantum Information*.

Quantum computers are currently in what is known as the noisy intermediate-scale quantum (NISQ) device era, where they have relatively few qubits, short circuit depth and low fidelity.[19] In this era, random physical errors in the hardware operations accumulate during computation and thus limit their performance and the types of computations they can perform. Consequently, to realise an advantage over classical computers, the programming of NISQ-era computers must be optimised for the computer's specific hardware architecture (analogous to the early days of classical computers). As different hardware architectures have different strengths and weaknesses, they are suited to different applications. Hence, the benchmarking of hardware architectures and the pursuit of optimal programming is critical in the NISQ era.

The types of applications where NISQ computers will likely deliver an advantage over classical computers are applications that can tolerate error, have sufficiently few inputs and outputs to be encoded by the limited number of qubits, yet involve sufficiently complex problems for the efficiency gains of quantum computation to be still beneficial. These applications are most likely found in signal/image processing, simulation and machine learning.

The next era for quantum computers is the error corrected (EC) era, where the number of qubits, circuit depths and fidelities have crossed thresholds that enable the implementation of methods that correct the random errors. In this era, programming no longer needs to be hardware-specific and quantum computers achieve their widest utility, including applications in optimisation, searching and factoring. Reaching the EC era is a significant technical challenge and is at least five years away for some hardware technologies and longer for others.

The future of quantum computing is one of diverse hardware that is integrated with classical computers throughout an information network. Different hardware technologies are suited to different roles in the network: centralised mainframe or massively parallelised cluster, distributed and edge. Those technologies that require cryogenic cooling or fragile control systems are better suited to mainframe computing roles in centralised cloud/supercomputing facilities; those that are more readily networked via quantum communications are better suited to forming distributed quantum computing networks; and those that have low size, weight and

---

19   For a critical discussion of the current state of quantum computing, see: Preskill, 'Quantum Computing in the NISQ Era and beyond'.

power and can be deeply integrated with classical computers are better suited to massive-parallelisation in centralised cloud/supercomputing clusters, integrated into distributed classical computers and deployed in mobile/edge computing devices. In all roles, quantum computers will be integrated with classical computers to some degree and act as an accelerator of specific computing tasks, akin to graphics accelerators of today. Thus, quantum computers should be not viewed as replacements for classical computers, but as augmentations and a continuation of the current trend towards computing hardware specialisation and diversification.

A key conclusion of this future vision of quantum computing is that a portfolio of complementary hardware technologies is required to fully exploit quantum computing across a defence information network in both the NISQ and EC eras. And that integration of hardware and software with classical computers is a critical pursuit.

Quantum computing has many potential defence applications. Select applications are:

a. *Enhanced signal and image processing and searching for intelligence, surveillance and reconnaissance (ISR).* Edge quantum computers potentially offer a leap in the ability to filter, decode, correlate or identify features in signals and images captured by edge ISR devices (e.g. satellites), thereby reducing the network congestion associated with streaming the raw data to centralised computers for processing. Distributed and centralised quantum computers may dramatically expand the capacity to search and extract intelligence from large unstructured databases.

b. *Enhanced optimisation of plans and logistics.* Quantum computers may be employed in headquarters at various levels to provide timely optimisation of complex plans and logistics systems.

c. *Enhanced AI/machine learning in automation, robotics and cyberwarfare.* Quantum computers promise to accelerate machine learning subroutines, learning and model optimisation, enabling enhanced performance and possibilities like in-mission relearning. The combination of enhanced signal/imaging processing, optimisation and machine learning may also improve decision-making in autonomous systems and robot populations, as well as new cyberwarfare tools.

d. *Enhanced operational simulation and geophysical modelling.* Quantum computers can accelerate the identification simulation of stochastic dynamics, like those that occur in operations, and systems described by complex differential equations, such as fluid dynamics in weather, ballistics and ocean dynamics. This improved simulation will support superior decision-making.

e. *Enhanced defence science and industry.* Quantum computers offer unprecedented capabilities in the simulation of complex molecules and materials, which will accelerate innovation in bio- and nanotechnology and materials engineering. As described above, quantum computers can also support enhanced engineering design and process optimisation, leading to higher-performing technologies and more efficient manufacturing.

f. *New cryptography capabilities.* As popularised, quantum computers have the ability to break public key encryption protocols using efficient algorithms for factoring large numbers. However, this requires many error-corrected qubits to achieve and is unlikely to occur within the next 10 years. As will be discussed in the next subsection, this timeframe is within the lifetime of some cryptographic systems (i.e. on airframes) and so the potential threat of quantum computers needs to be considered now when designing/specifying those systems. One cannot rule out that other quantum decryption algorithms will be discovered in the near future, and as a consequence, quantum computers have stimulated a new development race between decryption methods and encryption protocols.

Enablers and countermeasures

As discussed earlier, the performance of a quantum technology is determined by both its qubits and its classical control system and methods. Thus, the production and employment of high-performance quantum technologies are enabled by:

a. *Quantum foundries.* Quantum foundries synthesise quantum systems from source materials. For example, the synthesis of high-purity semiconducting and superconducting materials and atomic/nano-scale engineering of qubits as defects within or through heterostructures of the materials. Foundries require significant infrastructure, plant and know-how to produce high-performance quantum systems.

b. *High-performance electronic, optical, mechanical and thermal systems.* These form the control systems and isolation systems of quantum technologies. For example, high-frequency and precision microwave components and microcontrollers (i.e. similar to radar), high-resolution lasers and precise optical elements, ultra-high vacuum systems and cryostats. Global supply of such specialised and high-performance components is restricted.

c. *Nano/microfabrication facilities.* Such facilities are generally required to manufacture devices that integrate the control systems with the quantum systems, such as nano-optical, -electronic and -mechanical structures. Similar to foundries, they require significant infrastructure, plant and know-how.

d. *High-performance software stacks.* Stacks of multiple layers of software are required to operate and apply quantum hardware with the greatest effect. At the lower levels, this includes embedded programming of control systems, control optimisation and characterisation tools and compilers. At higher levels, this includes user interfaces, applications and interfaces with larger computing, communication and sensing systems. Development is currently occurring across all layers of the stack, with industry standards and architectures emerging, but not yet established.

e. *Benchmarking, testing and simulation facilities.* Quantum technologies are generally entering a phase in their development where independent and trusted benchmarking and testing facilities are required to accelerate development and support user adoption. For example, various public supercomputing facilities around the world have commenced programs to benchmark different quantum computing technologies. Such supercomputing facilities are also providing computing resources for the simulation of quantum technologies. Simulation is important to the identification, development and demonstration of applications in advance of the corresponding hardware maturing.

f. *Quantum-ready workforce.* The design, production and employment of quantum technologies require scientists and engineers of multiple disciplines. Some are necessarily quantum scientists and engineers, but many others must be experts who have converted from other fields (e.g. electronics, optics, software etc.) to bring their specialist skills and experience that are otherwise absent in the new industry.

These enablers define the critical supplies and infrastructure of a quantum industry.

There are a variety of possible countermeasures to quantum technology. Some are highly developed (e.g. attacks on QKD protocols), some are emerging (e.g. post-quantum cryptography) and others are yet to be (publicly) explored. Key examples are:

a. *Post-quantum cryptography.* The principal motivation for QKD is the potential threat posed by quantum computers to current RSA[20] public key encryption protocols by their ability to efficiently factor the large numbers used in the protocols, although it is unlikely that EC quantum computers with sufficient qubits to perform this task will be developed within the next 10 years. An alternate way of countering this threat is to switch RSA encryption protocols to new post-quantum encryption protocols that cannot be efficiently attacked by quantum computers. There are a variety of such protocols already available, and others are likely under development. As they generally require changes to the hardware and software of existing communication systems, their adoption needs to begin soon so that networks are 'quantum-resistant' before the advent of large EC quantum computers. Notably, post-quantum cryptography largely reduces the motivation for QKD. The remaining motivation is that there is always the risk that new algorithms are discovered for either quantum or classical computers that can efficiently attack the post-quantum encryption protocols.

b. *Disrupting quantum communications.* Methods for attacking the security of QKD protocols have been developed alongside the protocols themselves. These vary from injecting optical noise into the communication link to rerouting the link and impersonating the receiver. Alternatively, since sensitive photodetectors are required to detect the transmitted photons, QKD systems can potentially be simply disabled temporarily or permanently using high-power light pulses that overload or damage the photodetectors.

c. *Spoofing quantum sensors.* Quantum sensors are susceptible to interference from the environment and so can potentially be spoofed by carefully engineered emissions from structures or vehicles that need to be hidden. For example, the universal fundamental properties that quantum sensors use to achieve long-term stability mean that they can also be efficiently targeted by tuned spoofing signals that alter or disrupt the measurements of the sensors. It does not appear that such spoofing technology is being publicly developed.

---

20   'RSA' refers to the Rivest–Shamir–Adleman public key encryption system.

d. *Disabling quantum computers.* Quantum computers are complex, precisely calibrated and, in some cases, fragile machines. With adequate knowledge of their design, they can potentially be disrupted or disabled in various ways.

Greater attention needs to be applied to the development of countermeasures alongside quantum technologies and their applications.

Assessment and vision of quantum technologies in war

Assessment

The assessment of how quantum technologies will be employed in and impact warfare requires consideration of:

a. the dimensions of development risk, time to application and potential advantage

b. the diversity and early stage of quantum technology, where the precise judgement of the above dimensions is not possible and many applications are yet to be discovered

c. the current and future contexts and requirements of military operations, and how they may be modified by quantum technologies and their convergence with other emerging technologies.

With this in mind, an initial assessment of quantum technologies is:

a. *Quantum sensing and microscopy.* The most immediate technologies and applications in positioning, navigation and timing, detection of gravitational and magnetic anomalies and microwave spectroscopy are highly valuable and are lowest in risk. These will likely yield important capabilities of navigation in GPS-denied situations, enhanced detection of subsurface structures and vehicles, awareness and locating of electromagnetic emitters, and enhanced radar. The application of human–machine interfacing is highly valuable, especially its ability to transform the process of teaming information and autonomous systems; however, it is also very high risk and distant in time. The direct utility of the other applications and technologies in military operations is unclear.

b. *Quantum communications and cryptography.* The value of QKD to defence forces is questionable because of the advent of post-quantum cryptography methods that can be more easily integrated throughout established communication systems, the fact that QKD networks will likely be limited to a few high-priority links (e.g. at the operational and strategic levels) due to technical constraints for some time and that they may be easily disrupted. The longer-term possibilities of networked quantum devices and memories, and clock synchronisation, are highly valuable due to their ability to multiply the effects of other quantum and classical technologies, albeit also high in risk.

c. *Quantum computing and simulation.* Quantum computers likely have many highly valuable applications in warfare. The nearest term and lowest risk are in delivering enhanced image/signal processing, planning and autonomous systems. The more distant and higher-risk applications are in modelling, intelligence gathering, cyberwarfare and cryptography. Having said that, all forms of quantum computing remain relatively high in risk.

d. *Enablers and countermeasures.* Post-quantum cryptography is an important preparation for the possibility of quantum computers capable of cryptography and should be a priority. The other countermeasures appear to be relatively simple options to ensure some technology parity with quantum-enabled opposing forces and should be developed alongside quantum technologies. The enablers define the supply chains and infrastructure that are important to the sustainment of a defence force's future quantum capabilities.

Vision

Based on this assessment, a vision for the most likely cumulative evolution of quantum technologies over the next 20 years is:

a. *Near term (5–10 years).* Quantum navigation systems employed in major vehicles, vessels and airframes. Field-deployable quantum sensors used to image subsurface structures and vehicles. Quantum-enhanced timing in digital communication networks. Experimentation of QKD in operational and strategic level communications. Post-quantum encryption being introduced into communication systems. Mainframe quantum computers enhance planning in operational and strategic level headquarters. Strategic and operational level countermeasures for quantum communications.

b. *Intermediate term (10–15 years).* Quantum navigation systems employed in minor vehicles and airframes as well as autonomous vehicles. Air, sea and land quantum sensors used to image subsurface and concealed structures and vehicles. Networks of quantum spectrometers used to detect and locate electromagnetic emitters. Chip-scale analysers assisting medical and environmental monitoring. Synchronisation of quantum networked clocks for improved communications and sensor correlation. Regular use of QKD in operational and strategic level communications. Widespread use of post-quantum encryption in communication networks. Distributed and edge quantum computers enhancing signal/image processing in ISR and machine learning in autonomous systems. Spoofing countermeasures to quantum sensors being introduced to conceal major vehicles and structures.

c. *Far term (15–20 years).* Quantum spectrometers enhancing radar systems. Quantum networks of sensors used for large-area monitoring and enhanced detection. Experimentation with human–machine interfacing using quantum sensors. Secure quantum communications with centralised mainframe quantum computers. Edge and distributed quantum computers integrated throughout the information network. Quantum computers assist with modelling, intelligence gathering, cyberwarfare and cryptography. Cyberwarfare tools for disabling quantum computers ready.

The greatest risks and most dangerous possibilities for a defence force within this vision are:

a. advent of quantum computers capable of cryptography prior to the introduction of QKD networks or post-quantum encryption into the force's networks

b. adversarial forces prepared to deny GPS and have developed quantum-enhanced navigation prior to the force

c. widespread QKD networks prior to the force's development of effective countermeasures

d. adversarial forces control critical aspects of the quantum supply chain or global infrastructure

e. adversarial autonomous and intelligent systems significantly enhanced by quantum computers and sensors prior to the force's.

Conclusion

Defence forces are in an intensifying global competition to gain advantage through understanding, co-developing and exploiting quantum technologies in warfare. The purpose of this chapter was to assist defence professionals in understanding this emerging technology and how it is expected to impact war over the next 20 years, so that they can meet this challenge and prepare their defence forces.

Accordingly, the key conclusions and implications of this chapter for defence professionals include, firstly, that quantum technologies are no longer scientific speculation. They are advancing rapidly and will begin to impact warfare in the next 5 years. Secondly, quantum technologies and their applications will also diversify, refine and become increasingly networked/ network-centric over the next 20 years. Countermeasures should be advanced alongside quantum technologies in order to ameliorate the risk of adversaries developing early advantages and as a tool to test and increase the resilience of quantum technologies. The enablers of quantum technologies are national strategic assets (i.e. foundries, facilities, workforces), and a strategic approach is required to secure sovereign capability and capture control of parts of the global supply chain. Finally, quantum technologies should be considered in the context of other emerging technologies. It is likely that their largest impact will be achieved through accelerating and/or enhancing these other technologies.

However, this chapter is only a beginning. It is likely that the most disruptive and highest-value quantum technologies and applications were not identified here and that the field of quantum technologies will quickly evolve. Thus, it is recommended that defence professionals take a portfolio approach that initially focuses on the scoping, discovery and validation of the applications of different quantum technologies, before committing resources to supporting the development of the most advantageous. Simultaneously, defence professionals should continuously curate an understanding of the global quantum technology landscape and national strategies to ensure national control and capabilities in the production, sustainment and employment of the technologies.

References

Arute, Frank, Kunal Arya, Ryan Babbush, Dave Bacon, Joseph C. Bardin, Rami Barends, Rupak Biswas, et al. 'Quantum Supremacy Using a Programmable Superconducting Processor'. *Nature* 574 (2019): 505–510.

Boto, Elena, Sofie S Meyer, Vishal Shah, Orang Alem, Svenja Knappe, Peter Kruger, T Mark Fromhold, et al. 'A New Generation of Magnetoencephalography: Room Temperature Measurements Using Optically-Pumped Magnetometers'. *NeuroImage* 149 (2017): 404–414. doi.org/10.1016/j.neuroimage.2017.01.034.

Cartlidge, Edwin. 'Quantum Sensors: A Revolution in the Offing?' *Optics & Photonics News* (September 2019): 24–31. doi.org/10.1364/OPN.30.9.000024.

Degen, CL, F Reinhard, and P Cappellaro. 'Quantum Sensing'. *Reviews of Modern Physics* 89 (2017): 035002. doi.org/10.1103/RevModPhys.89.035002.

Department of Defence. *Army Quantum Technology Roadmap.* Canberra: Australian Government Publishing Service, 2021.

Doherty, Marcus W. 'Quantum Microscopy'. *AOS News* 30 (2016): 2.

Doherty, Marcus W. 'Quantum Technology: An Introduction'. *Land Power Forum, Australian Army Research Centre,* 27 May 2020, researchcentre.army.gov.au/library/land-power-forum/quantum-technology-introduction.

Doherty, Marcus W, Andre Luiten, David Simpson, and Liam Hall. 'Quantum Technology: Sensing and Imaging'. *Land Power Forum, Australian Army Research Centre,* 2 September 2020, researchcentre.army.gov.au/library/land-power-forum/quantum-technology-sensing-and-imaging.

Dowling, Jonathan P, and Gerard J Milburn. 'Quantum Technology: The Second Quantum Revolution'. *Philosophical Transactions of the Royal Society A* 361, no. 1809 (2003): 1655. doi.org/10.1098/rsta.2003.1227.

Gambetta, Jay. 'IBM's Roadmap for Scaling Quantum Technology'. *IBM Research Blog,* 15 September 2020. www.ibm.com/blogs/research/2020/09/ibm-quantum-roadmap/.

Gibney, Elizabeth. 'Quantum Gold Rush: The Private Funding Pouring into Quantum Start-Ups'. *Nature* 574, no. 7776 (2019): 22. doi.org/10.1038/d41586-019-02935-4.

Kelly, Julian. 'A Preview of Bristlecone, Google's New Quantum Processor'. *Google AI Blog,* 5 March 2018. ai.googleblog.com/2018/03/a-preview-of-bristlecone-googles-new.html [site discontinued].

Nielsen, Michael A, and Isaac L Chuang. *Quantum Computation and Quantum Information*. 10th anniversary edition. Cambridge: Cambridge University Press, 2010.

Palmer, Jason. 'Here, There and Everywhere: Technology Quarterly'. *The Economist*, 11 March 2017.

Preskill, John. 'Quantum Computing in the NISQ Era and beyond'. *Quantum* 2 (2018): 79, doi.org/10.22331/q-2018-08-06-79.

Susskind, Leonard, and Art Friedman. *Quantum Mechanics: The Theoretical Minimum*. New York: Basic Books, 2015.

Vedral, Vlatko. 'Instant Expert 33: Quantum Information'. *New Scientist*, multiple dates, www.newscientist.com/round-up/quantum-information/.

Wehner, Stephanie, David Elkouss, and Ronald Hanson. 'Quantum Internet: A Vision for the Road Ahead'. *Science* 362, no. 6412 (2018): eaam9288. doi.org/10.1126/science.aam9288.

8

The Distance Paradox: Reaper, the Human Dimension of Remote Warfare and Future Challenges for the Royal Air Force

Peter Lee[1]

Introduction[2]

In the final decade of the Royal Air Force's (RAF) first hundred years of existence, the addition of the MQ-9 Reaper to its inventory has brought both operational benefits[3] and public controversy.[4] In 2007 the Reaper was introduced as an urgent operational requirement for Operation Herrick

1   I would like to thank Alfie Macadam for his assistance in literature searching during his research placement week with me in June 2018.
2   Editors' note: as mentioned in the Introduction, this chapter was written in 2020, and global events since then may have overtaken some aspects. As the pace of technological and cultural change has continued to accelerate, the editors have opted to present this text as drafted to minimise further delays in publication.
3   House of Commons Defence Committee, *Remote Control: Remotely Piloted Air Systems—Current and Future UK Use: Government Response to the Committee's Tenth Report of Session 2013–14*, Sixth Special Report of Session 2014–15 (London: House of Commons, 20 July 2014), 14, publications.parliament. uk/pa/cm201415/cmselect/cmdfence/611/611.pdf.
4   Ben Farmer, 'RAF Reaper Drone Interrupts Islamic State Public Execution', *The Telegraph*, 16 May 2017, www.telegraph.co.uk/news/2017/05/16/raf-reaper-drone-interrupts-islamic-state-public-execution/; BBC, 'Syria War: MoD Admits Civilian Died in RAF Strike on Islamic State', *BBC*, 2 May 2018, www. bbc.co.uk/news/uk-43977394.

in Afghanistan.[5] By 2014, the Reaper's dual intelligence, surveillance and reconnaissance (ISR) and attack capabilities had proved to be so effective and indispensable that it was brought into RAF Core Capability.[6] To staff this new capability, personnel were brought together from a range of flying backgrounds, including the Harrier, Tornado F3 and GR4, Nimrod, Hercules and various helicopters—and some had no flying background. Individuals were drawn from the RAF, Royal Navy, Army and Royal Marines to develop and apply the Reaper's ability to remotely deliver air power effect in counterinsurgency operations, initially in Afghanistan and latterly in Iraq and Syria. The disparate backgrounds and experiences of Reaper personnel have resulted in squadron cultures that have often reflected the dominant previous experiences and personal ethos of those involved.

The purpose of this chapter is to reflect on key aspects of the human operator dimension of remote air operations through the experiences of Reaper personnel, linking current challenges to historical precedent, and identifying future challenges that will need to be addressed to optimise performance and resilience in the decades to come as the Reaper is eventually succeeded by the Protector.[7] This article forms part of a wider study entitled, 'Royal Air Force Reaper: 21st Century Air Warfare from the Operators' Perspective',[8] the major output of which was a book entitled *Reaper Force: Inside Britain's Drone Wars*.[9] Empirical, qualitative data were collected between July 2016 and February 2018, comprising observational field research with Nos XIII and 39 squadrons and 90 semi-structured interviews with members of the RAF Reaper community.[10] In the course of the data gathering, recurring factors that could contribute to sustained, long-term participation in operations, otherwise referred to as 'resilience', were identified. From this coding of notes and observations emerged the following Reaper Resilience Matrix:

---

5   HC 772 Defence Committee, 'Written Evidence from the Royal Aeronautical Society', *parliament. uk*, 15 September 2013, para. 24, publications.parliament.uk/pa/cm201314/cmselect/cmdfence/772/772vw13.htm.

6   House of Commons Defence Committee, *Remote Control*, 15.

7   Ministry of Defence, 'New Investment in Counter Terrorism for UK Armed Forces', *GOV.UK*, 4 October 2015, www.gov.uk/government/news/new-investment-in-counter-terrorism-for-uk-armed-forces.

8   'Royal Air Force Reaper: 21st Century Air Warfare from the Operators' Perspective', University of Portsmouth Research Ethics Committee Protocol E365, approved 21 October 2015, Ministry of Defence Research Ethics Committee Protocol 707/MODREC/15, approved 1 July 2016.

9   Peter Lee, *Reaper Force: Inside Britain's Drone Wars* (London: John Blake Publishing, 2018).

10   Ninety interviews were conducted as follows: 45 current crew members (at the time of interview); 21 former crew members from both squadrons; 24 spouses and partners. The military participants included both women and men and ranged in non-commissioned officer rank from corporal to warrant officer, and in officer rank from flight lieutenant to wing commander.

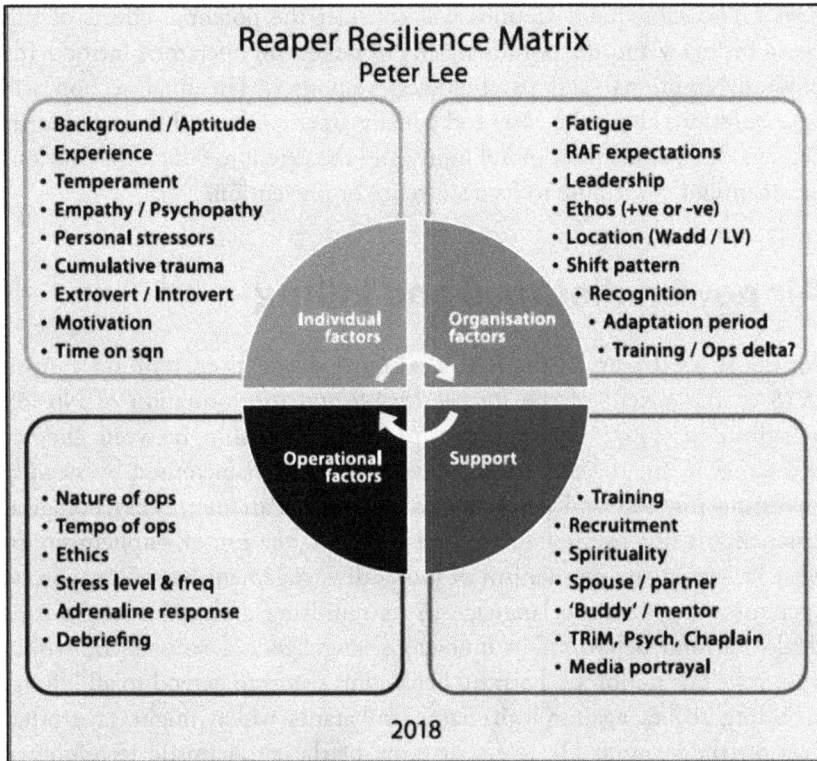

Figure 8.1: Reaper Resilience Matrix.
Source: Author's summary.

There is insufficient scope within this chapter to address every element of this matrix. However, several of the factors identified here will be explored below to understand some current and future challenges in the human dimension of RAF remotely piloted air operations.

The first section will provide a brief historical overview showing how physical and psychological distance in weapon use by air, land and maritime forces increased in tandem over time.[11] Snipers, special forces, fast jet crews, intelligence personnel and others have experienced elements of this dynamic. However, this study focuses on remotely piloted aircraft and their crews: a point raised by Killeen and Jordan in 2013 and expanded upon

---

11   A good starting point for further study is Dave Grossman, *On Killing: The Psychological Cost of Learning to Kill in War and Society*, revised edition (New York: Back Bay Books, 2009).

here.[12] The subsequent sections will consider the potential effects of this visual intimacy and the re-humanising of targets on operators, introducing physical, emotional and psychological responses. The final section will pose questions about the costs and benefits of empathy with human targets before a discussion about moral injury and the extent to which the previous factors might contribute to its occurrence or prevention.

Air power, distance and killing

For the first 90 years of the RAF's existence, from its creation on 1 April 1918 to the advent of the Reaper MQ-9 and the formation of No. 39 Squadron in 2007, the trajectory of the relationship between shooter and target in the delivery of air power has been characterised by steadily increasing physical and psychological distance. Part of the psychological distancing is, for psychologist Albert Bandura, the use of 'euphemism' in what he sees as the mechanism of moral disengagement by military drone operators: 'Euphemistic language in its sanitising and convoluted forms cloaks harmful behaviour in innocuous language and removes humanity from it.'[13] His notion of 'harmful behaviour' seems to extend to all killing, including strikes against legitimate combatants which might save other lives on the ground. He refers to some of the euphemistic terminology used in the US drone community: 'touchdowns … jackpots … personality strikes … signature strikes', as well as 'collateral damage' (the unintended, accidental or indirect killing of civilians), the latter also being used by RAF Reaper personnel.[14] Bandura's argument about the use of euphemism is, to a significant extent, echoed in the empirical research with military veterans conducted by Grossman.[15] The practical result of the use of euphemistic language is the emotional distance between the shooter and the human target.

The reason for this use of euphemism in the delivery of air power, or in any other element of killing in war or armed conflict, is to help facilitate an activity—killing—that is difficult for even many hardened military personnel. Grossman points out that 'there is within most men [and women]

---

12   Damian Killeen and David Jordan, 'RPAS: Future Force or Force Multiplier? An Analysis of Manned/Unmanned Platforms and Force Balancing', *Air Power Review* 16, no. 3 (2013): 22–23.
13   Albert Bandura, 'Disengaging Morality from Robotic War', *The Psychologist* 30 (February 2017): 39.
14   Bandura, 'Disengaging Morality', 41.
15   Grossman, *On Killing*.

an intense resistance to killing their fellow man [or woman]'. He goes on to observe that 'throughout history the majority of men on the battlefield would not attempt to kill the enemy, even to save their own lives or the lives of their friends'.[16] In his study, he presents evidence to demonstrate that only a small minority of combatants—throughout history and up to the present—are comfortable, or at ease, with killing, even in a life-threatening wartime scenario:

> The burden of killing is so great that most men try not to admit that they have killed. They deny it to others, and they try to deny it to themselves.[17]

That trajectory of increasing distance in the air power 'killing' domain precedes the RAF, starting with the Royal Flying Corps and Royal Naval Air Service in the First World War. The recollections of Major James McCudden VC provide a good indication of the distances involved in air-to-air combat a century ago. On 13 January 1918 he was flying behind enemy lines at 17,000 feet when he spotted a two-seater enemy aircraft a few thousand feet below.[18] He idled his engine and, almost silently, glided down behind his unsuspecting target at 9,000 feet:

> when I got within good close range, about 100 yards, I pressed both triggers; my two guns responded well, and I saw pieces of three-ply wood fall off the side of the Hun's fuselage. Then the L.V.G. went into a flat, right-hand spiral glide until it hit the ground a mass of flying wreckage … I hate to shoot the Hun down without him seeing me, for although this method is in accordance with my doctrine, it is against what little sporting instincts I have left.[19]

Two separate indicators of distance are captured in these few words: physical distance and psychological distance. At the time of firing, McCudden was about 100 yards physically distant from the German aircraft he shot down. Then, afterwards, he watched the downed enemy aircraft hit the ground. Assuming that McCudden maintained his altitude at 9,000 feet, to give him an advantage in any subsequent air-to-air encounter, he would be watching from almost 3,000 metres vertically away from the crash site. In addition, psychological distance is indicated in two ways. First, he refers euphemistically to 'the Hun', rather than the 'pilot' or 'crew' or 'people';

---

16   Grossman, *On Killing*, 4.
17   Grossman, *On Killing*, 91.
18   James B McCudden, *Flying Fury* (Poland: Amazon Fulfilment, 2009), 219.
19   McCudden, *Flying Fury*, 219.

second, he describes 'a mass of flying wreckage', focusing on what happened to the aircraft rather than on the deaths of the crew. A few days later, after shooting down another German aircraft, McCudden described:

> This D.F.W crew deserved to die, because they had no notion whatever of how to defend themselves, which showed that during their training they must have been slack, and lazy, and probably liked going to Berlin too often instead of sticking to their training and learning as much as they could while they had the opportunity. I had no sympathy for those fellows, and that is the mental estimate which I formed of them while flying back to my aerodrome to report the destruction of my 43rd aerial victim.[20]

In this description, McCudden reduces the humanity of the crew he shot down, discursively imagining a pairing who 'deserved to die' for being 'slack', 'lazy' and too keen on having a good time in Berlin rather than training properly. One possible alternative explanation for McCudden's choice of words is that he was justifying their killing *to himself* on some psychological level and avoiding a possible alternative explanation: inexperienced crew with little or no chance against one of the RFC's—later RAF's—most effective pilots.

By the Second World War and afterwards, the dynamic of killing from the air had moved on considerably. That increasing physical distance, and accompanying sanitising language, can be seen in Group Captain Leonard Cheshire's recollections from a bombing raid against Cologne:

> If what we saw below was true, Cologne was destroyed … Cologne was burning, it was burning as no city in the world can ever have burnt, and with it was burning the morale of the German citizen.[21]

There was no possibility of seeing individuals from his bombing altitude even as Cheshire referred to the burning city below. His reference was to the burning of the *morale* of the German citizens and not to the burning of the German citizens themselves. Cheshire's use of language was consistent with the language of the official bombing directives issued to Bomber Command in the Second World War. A directive on 14 February 1942 to Acting Air Officer Commanding-in-Chief (AOC-in-C) Bomber Command, JEA Baldwin stated: 'the primary object of your operations should now be focussed on the morale of the enemy civilian population and in particular,

---

20   McCudden, *Flying Fury*, 224.
21   Leonard Cheshire, *Bomber Pilot* (London: HarperCollins, 1975), 135.

of the industrial workers'.[22] The directive does not demand that civilians themselves are targeted, merely that their *morale* is undermined: through the bombing (including incendiaries) of their homes and neighbourhoods in the way that Cheshire observed. Garret argues that the Air Ministry maintained 'and sustained [a] public fiction about Bomber Command's strategy' through vague references to undermining morale.[23] In October 1943 Arthur Harris, AOC-in-C, Bomber Command, encouraged greater candour about the reality of bombing and urged the Air Ministry to 'stop their public denials that the bombing campaign was focussed on "the obliteration of German cities and their inhabitants as such"'.[24] The use of language to either describe or obscure the activities and aims of Bomber Command was a point of sustained debate until the end of the war.

Over the same period, American political and military leaders similarly sought to deploy language in a way that presented their actions more favourably to the public. Hays Parks points out that there was little evidence that the US Army Air Force was more accurate with its 'precision', 'pickle barrel', and 'pin-point' bombing than the RAF's area offensive being conducted against Germany at the time.[25] There is, however, evidence of two things that the British and Americans had in common in their respective bombing campaigns. First, bombing altitudes were getting higher (25,000 feet),[26] thereby increasing the physical distance of the bomber crews from those they killed beyond that experienced by First World War aircrew. Second, the use of sanitising language, like 'area bombing', 'morale bombing' or 'carpet bombing' psychologically protected those involved from confronting the stark reality of killing civilians over entire city areas.[27] The use of language was a factor in emotionally distancing political leaders, military commanders and bomber crews—and the general public—from the 'killing' aspect of the bombing policies they undertook.

---

22 Charles Webster and Noble Frankland, *The Strategic Air Offensive against Germany 1939–1945 Vol. IV* (London: Her Majesty's Stationery Office, 1961), 144, Directive on 14 February 1942.

23 Stephen A Garrett, *Ethics and Airpower in World War II: The British Bombing of German Cities* (New York: St. Martin's Press, 1993), 32.

24 Garrett, *Ethics and Airpower*, 32–33.

25 W Hays Parks, '"Precision" and "Area" Bombing: Who Did Which, and When?' *Journal of Strategic Studies* 18, no. 1 (1995): 146–147.

26 Hays Parks, '"Precision" and "Area" Bombing', 148.

27 I have written elsewhere about the ethics of the German city bombings and explored the nuances of 'intention', unintended consequences and so on. There is not the scope here to reprise those arguments. See Peter Lee, 'Return from the Wilderness: An Assessment of Arthur Harris's Moral Responsibility for the German City Bombings', *Air Power Review* 16, no. 1 (Spring 2013): 70–90; Peter Lee and Colin McHattie, 'Churchill and the Bombing of German Cities 1940–1945', *Global War Studies* 13, no. 1 (2016): 47–69.

In the decades following the Second World War, the distances involved in the delivery of air power—especially air-to-ground attacks—continued to increase. In parallel, aircraft got faster and bombing runs took correspondingly less time. By 2003, the RAF had acquired the Storm Shadow stand-off cruise missile for use in Operation Telic. It could be fired from more than 500 km away, typically against command and control centres, airfields, communications hubs, ammunition storage and other key targets.[28] The Storm Shadow would officially enter RAF service in 2004, increasing the distance between aircrew and target—its 'fire and forget' preprogrammed capability providing parallel psychological distance from targets for those involved. At the same time, the US Predator remotely piloted aircraft program was rapidly expanding—with embedded RAF personnel involved—placing aircrew thousands of miles away from not only their targets but from their aircraft as well. Then in 2007, the RAF re-formed No. 39 Squadron with the MQ-9A Reaper, officially joining the era of remotely piloted air operations.[29] The distance between the shooter and target would appear to have reached new levels.

Reaper and the distance paradox

A number of now-familiar tropes emerged in the 2000s in popular and academic critique of military drones like the Reaper. The 'Playstation mentality' meme is probably the best known of those representations of Reaper crews, with the claim that:

> Geographical and psychological distance between the drone operator and the target lowers the threshold in regard to launching an attack … Operators, rather than seeing human beings, perceive mere blips on a screen.[30]

Singer, similarly, argued that drone crews are 'disconnected' from the wars in which they conduct air operations.[31] Further, Olsthoorn referred to both the psychological and physical distance from their targets, as though the

---

28   Gerald Wright, 'The Storm Shadow Cruise Missile', *UK Defence Journal*, 20 October 2015, ukdefence journal.org.uk/the-storm-shadow-cruise-missile/.

29   Royal Air Force, 'Reaper (MQ-9A)', *Royal Air Force*, www.raf.mod.uk/aircraft/reaper-mq9a/, accessed 15 June 2018.

30   Chris Cole, Mary Dobbing, and Amy Hailwood, *Convenient Killing: Armed Drones and the 'Playstation' Mentality* (Oxford: Fellowship of Reconciliation, 2010), 4.

31   Peter W Singer, *Wired for War* (New York: Penguin, 2009), 332.

two are directly linked,[32] while Benjamin perpetuated the assumption that killing by drone strike from afar was somehow 'easier' than conventional military killing.[33] These authors all appear to have assumed that the historical link between physical and psychological distance continued with US and UK remotely piloted air operations using the Predator and the Reaper. Further, limited primary research restricted authors' understanding of the acuity of the images that could be seen on the screens, and hence the mentally immersive nature of such operations. In contrast, in 2014, a report by the House of Commons Defence Committee acknowledged the All Party Parliamentary Group on Drones' concerns about the 'limited consideration of the psychological impact of drones *on operators* and those living in affected areas'.[34] However, public focus, and academic and media enquiry, concentrated on policy surrounding remotely piloted aircraft systems (RPAS) use and on those in affected areas, rather than on the operators.[35]

However, over that same period, the impact of remote operations—especially killing—on the crews who carry them out has quietly but steadily grown as an area of interest and enquiry. In 2013, American former Predator pilot Brandon Bryant publicly shared his diagnosis of post-traumatic stress disorder (PTSD), becoming something of a *cause celebre* in describing multiple kills—with apparent scant regard for the deaths of civilians—and the extremes of behaviour that he experienced outside the Ground Control Station.[36] More recently, US Air Force (USAF) Imagery Analysts have been reported as experiencing varying degrees of mental trauma, and 'nearly one in five had witnessed a rape within the past year. Some airmen reported witnessing more than 100 incidents of rape or torture' according to USAF Wing Surgeon Lieutenant Colonel Cameron Thurman.[37] Then, on 13 June 2018, the *New York Times* published a major investigative piece on the

---

32   Peter Olsthoorn, *Military Ethics and Virtues: An Interdisciplinary Approach for the 21st Century* (New York: Routledge, 2011), 126.

33   Medea Benjamin, *Drone Warfare: Killing by Remote Control* (New York and London: OR Books, 2012).

34   House of Commons Defence Committee, *Remote Control*, 42 (italics added for emphasis).

35   James Cavallaro, Stephan Sonnenberg and Sarah Knuckey, *Living under Drones: Death, Injury and Trauma to Civilians from US Drone Practices in Pakistan* (September 2012), International Human Rights and Conflict Resolution Clinic (Stanford Law School) and the Global Justice Clinic (NYU School of Law), law.stanford.edu/publications/living-under-drones-death-injury-and-trauma-to-civilians-from-us-drone-practices-in-pakistan/.

36   Matthew Power, 'Confessions of a Drone Warrior', *GQ*, 22 October 2013, www.gq.com/story/drone-uav-pilot-assassination?currentPage=4.

37   Sarah McCammon, 'The Warfare May be Remote but the Trauma Is Real', *NPR*, 24 April 2017, www.npr.org/2017/04/24/525413427/for-drone-pilots-warfare-may-be-remote-but-the-trauma-is-real.

mental trauma experienced by US drone crews, asking whether some or all of these individuals could be suffering from 'moral injury'?[38] 'Moral injury' will be discussed below but an initial sense of this contested term is found in Nash et al.'s definition, which refers to 'changes in biological, psychological, social, or spiritual functioning resulting from witnessing or perpetrating acts or failures to act that transgress deeply held, communally shared moral beliefs and expectations'.[39]

Against this backdrop, the distance paradox experienced by Reaper personnel emerged right away in my research interviews, in their descriptions of events they had witnessed on the ground via the sensor suite on the aircraft. They were physically located at either Creech Air Force Base in Nevada or at RAF Waddington in Lincolnshire. Aircraft crews had never been so geographically far away from their targets, yet they witnessed and experienced events on the ground in great detail. In addition, those events were juxtaposed with the banalities of day-to-day family life:

> I am a parent governor for my local school and every year I volunteer to go away with the teaching staff and help the kids enjoy the great outdoors. It's only 3 days away but the kids get to abseil, canoe, pot-hole and do many other fun things. One year, I had a great time and thoroughly enjoyed the company of the children and the staff. Eighteen hours after I got back I was in work, watching a prisoner having his head cut off and being powerless to do anything about it. Oh how my life had changed—and not for the better—in such a short period of time! (Simmo—Sensor Operator)[40]

Comments like this and many others made it clear, subsequently reinforced over time, that the incidents which had the greatest impact on the Reaper crew members—not surprisingly—revolved around killing or serious physical harm. These either involved their own use of weapons and killing (or seriously harming) enemies on the ground, or having to watch as atrocities—including beheadings or other executions—were perpetrated, while powerless to intervene. At maximum camera resolution, the view the Reaper crews have of these events is now not much different to the 100 yards

---

38 Eyal Press, 'The Wounds of the Drone Warrior', *New York Times*, 13 June 2018, www.nytimes.com/2018/06/13/magazine/veterans-ptsd-drone-warrior-wounds.html.

39 William P Nash et al., 'Consensus Recommendations for Common Data Elements for Operational Stress Research and Surveillance: Report of a Federal Interagency Working Group', *Archives of Physical Medicine and Rehabilitation* 91, no. 11 (2010): 1677.

40 Lee, *Reaper Force*, 266–67. Pseudonyms are used throughout in accordance with my Research Ethics protocol. Ranks are omitted unless relevant to the point being made, and no participants asked for their gender to be obscured or anonymised.

or so distance between aerial combatants in the First World War described previously. More significant is that First World War aircrew who shot down enemy aircraft were visually further away when the target aircraft crashed on the ground than are observers of a Reaper full motion video feed today.

Blair and House say of this phenomenon in remote air operations:

> We hold that the operative distance is not physical distance, but cognitive distance. For remote warriors, Cognitive Combat Intimacy (CCI) is a relational attachment to a human target mediated by sensor resolution and dwell time, or duration of observation.[41]

This concept of Cognitive Combat Intimacy has both advantages and disadvantages. Positively, it recognises the mental engagement of the Reaper pilot, sensor operator and mission intelligence coordinator. It also encapsulates the cognitive intimacy of the authorising officer and senior mission intelligence coordinator, imagery analysts in intelligence agencies, and commanders with a live video feed. Potentially more negatively, that intimacy is somewhat removed from the constant distance–intimacy paradox at the heart of remote warfare. The paradox itself is a factor in creating, for some, a cognitive dissonance between the physical distance and safety enjoyed by the Reaper crew, and the emotional intimacy and psychological threat. In the example of 'Simmo' above, his physical distance from a war zone enabled him to pursue relaxing educational and family activities as a school governor, while on his return to work the Reaper full motion video feed brought him great visual and psychological intimacy with a beheading victim.

Grossman observes:

> During strategic bombing [in the Second World War] the pilots and bombardiers were protected by distance and could deny to themselves that they were attempting to kill any specific individual.[42]

Geographical distance and the security of a Ground Control Station provide the ultimate physical protection for Reaper crew members. However, the clarity and persistence of the close-up views they see do not afford them the denial and psychological protection that Grossman argues the bomber crews

---

41   Dave Blair and Karen House, 'Avengers in Wrath: Moral Agency and Trauma Prevention for Remote Warriors', *Lawfare*, 12 November 2017, www.lawfaremedia.org/article/avengers-wrath-moral-agency-and-trauma-prevention-remote-warriors. This paper is highly recommended reading for the detail of its psychological insights, which are not addressed here.
42   Grossman, *On Killing*, 78.

experienced in the Second World War. Crucially, significant time and effort goes into ensuring that *specific, identified individuals* are killed. Days, weeks and even months have gone into observing particular high-value targets in Afghanistan, Iraq and Syria.

Historically, an important means of enabling the killing of enemies is through psychological distancing and dehumanisation, and there is extensive literature on the phenomenon. At the extreme end of this 'dehumanising' spectrum are atrocities against the Jews in the Second World War, which 'originates from the delegitimisation of the Jews by the Nazi regime',[43] and the Cambodian genocide in the 1970s. The mass killing of innocents— or non-combatants, to use a less loaded term—requires a high degree of dehumanisation which strips the victim of selfhood, identity, culture and intrinsic value.[44]

In his use of language, McCudden's words (quoted earlier in this chapter) conform to a pattern of behaviour and use of language in relation to killing in war. Bandura states: 'By camouflaging pernicious activities in innocent or sanitizing parlance, the activities lose much of their repugnancy. Soldiers "waste" people rather than kill them.'[45] Hence, a century ago McCudden described watching the German LVG aircraft going into a spin and crashing, rather than writing about killing the crew. Further: 'Self-censure for cruel conduct can be disengaged by stripping people of human qualities … They are portrayed as mindless "savages," "gooks," and other despicable wretches.'[46] Or, in McCudden's case—consistent with the language and attitudes in that particular war at that time—he shot down 'the Hun'.

If it is even partially correct that combatants have previously dehumanised their enemies in order to kill them, the approach often breaks down with the visual intimacy of killing from a Reaper. One Reaper pilot—a highly experienced former fast jet pilot—describes some of the dynamics involved in the humanising of a potential target, and the visual and psychological intimacy it entails:

---

43   Chaiara Volpato and Alberta Contarello, 'Towards a Social Psychology of Extreme Situations: Primo Levi's *If This Is a Man* and Social Identity Theory', *European Journal of Social Psychology* 29 (1999): 252.

44   Herbert G Kelman, 'Violence without Moral Restraint: Reflections on the Dehumanization of Victims and Victimizers', *Journal of Social Issues* 29, no. 4 (Fall 1973): 25–61.

45   Albert Bandura, 'Moral Disengagement in the Perpetration of Inhumanities', *Personality and Social Psychology Review* 3, no. 3 (1999): 195.

46   Bandura, 'Moral Disengagement', 200.

We may watch 'Target A' for weeks, building up a pattern of life for the individual: know exactly what time he eats his meals; drives to the Mosque; or uses the ablutions—outdoors of course! This is all-important for the guys on the ground. However, what we also see is the individual interacting with his family—playing with his kids and helping his wife around the compound. When a strike goes in, we stay on station and see the reactions of the wife and kids when the body is brought to them. You see someone fall to the floor and sob so hard their body is convulsing. A conventional aircraft often doesn't have the endurance [in the air] to witness this.[47]

The level of visual detail afforded the Reaper crew, and the sustained surveillance involved in this example, does not allow the individuals involved to deny that the enemy has a family or a fully rounded life. If part of that life is devoted to insurgency warfare, then it is viewed in a broader context. Those involved not only see the actual killing in detail, they also see the immediate consequences: recovery of a body or body parts, family reactions, funeral and so on. Such potentially traumatic visual stimuli are not unique to the Reaper Force but there is not the scope here to explore the similar experiences of online sex-crime police investigators, war photographers, defence intelligence analysts, or Facebook and YouTube online moderators.

Potential impacts of remote warfare

If Bandura is correct, the visual and emotional intimacy of Reaper operations will bring its own psychological challenges because it is 'difficult to mistreat humanized persons without suffering personal distress and self-condemnation'.[48] Before accepting his argument, however, the word 'mistreat' should be qualified. In the context of military operations, it may be the case that killing another human being is an appropriate operational and ethical act that can result in personal and professional satisfaction. Such a response does not preclude the possibility of sadness or regret that a child may have been left without a parent who happened to be an enemy combatant. Responses to the taking of life in war are complex and individualised. Having spent so much time with so many Reaper personnel, whose responses to conducting remote operations ranged from apparently unaffected to significantly affected—with the majority somewhere in

47   Lee, *Reaper Force*, 170.
48   Lee, *Reaper Force*, 170.

between—a new question emerged: ***Why are some crew members able to operate for five, six or even seven years consecutively, while others seem exhausted after two?*** I will go as far as to suggest that the answer to this question—and only the beginning of an answer is offered in this chapter—will shape the human dimension of remote air operations, and therefore the culture of the RAF itself, long into the future.

The most consistent response to how long a Reaper crew member could or should operate without an extended break, came in the interviews with spouses and partners: somewhere between 2.5 and 3 years.[49] Vet 18, a former Reaper Squadron Commander, had reflected on this question for years and concluded that, although the answer would always be individualised, 'None will *happily* make it to the end of 2 tours.'[50] Underpinning almost every discussion with operators and spouses/partners about longevity on the Reaper Force was fatigue: a constant tiredness that was rooted in long days on a '6 days on, 3 days off' work pattern that rarely worked out so neatly or generously. 'He's tired *all* the time', says Partner 20, a phrase that is repeated so often during interviews it could be the unofficial Reaper Force motto. But in the background of the 'fatigue' discussions, a common link to operations and weapon use also emerged—especially where human targets were involved. Partner 19 tried to explain his wife almost collapsing, in tears, onto the floor when she got home after an extended, intense period of Reaper activity: 'I put it partly down to exhaustion and partly down to the operation she was conducting that day.' Partner 12 echoed those sentiments, while also explaining the reason her husband persevered:

> the physical impact on his body, lack of sleep, constant mental exhaustion—just not getting 'down time'. That's what I worried about most. But I don't think he thought about doing anything else—that was just his job.

The role itself, for the vast majority of Reaper personnel, is as professionally and personally fulfilling as it is demanding.

---

49 Methodology note: these were consistent but *qualitative* observations made in interviews with 25 spouses and partners, and not rigorous, quantitative measures.
50 About five years in total.

Shooting, killing and physical response

The physiological effects of conducting lethal missile or bomb strikes is one area where the experiences of Reaper crew members consistently contradict public claims that they are somehow emotionally distant, remote game-players. Almost all crew members described adrenaline spikes—sometimes almost overwhelmingly powerful—and rapid heart rates in the build-up to, and execution of, a weapon strike, though for some these lessened over time. In July 2016 I observed my first lethal missile strike in real time in a 39 Squadron Ground Control Station, while sitting immediately behind the sensor operator as he guided a Hellfire missile onto his human target: an ISIS (Islamic State in Iraq and Syria) fighter on a motorbike. As the missile struck, hitting and killing the target, the sensor operator—who had been holding his breath for the half-minute duration of the missile flight—exhaled with relief and exclaimed: 'My heart's beating out of my chest!'[51] That link between heart rate, adrenaline spike and weapon firing is highly prevalent among crew members and can be especially tense for those new to shooting and killing.

Jeff, an RPAS(P),[52] was building up to his first missile strike, just a couple of days after he had qualified as Combat Ready: 'My heart has never gone so fast. I've done a lot of silly things in my time that has raised my heart rate quite considerably, but not like this.' However, actual weapon firing—depending on the operational environment, rules of engagement (ROE) and proximity of non-combatants—might only happen once for every five or six times the pilot starts the process of getting all the authorisations for a strike, or is called in for a potential strike. For most crew members, just beginning the process prompts a spike in adrenaline. Ross gives a detailed insight and adds a further dimension to the personal dynamics in a weapon strike:

> When you've had a 'decent' strike event you go through a process of extreme adrenaline. It gets to a peak … you ride the crest, the strike happens, then you come down as the adrenaline subsides. The occasions when a strike doesn't happen—maybe because there

---

51 Lee, *Reaper Force*, 59.
52 RPAS(P): an individual who was recruited to the Reaper Force and trained specifically as a Reaper pilot, as opposed to a pilot of another aircraft type who transferred across.

> are civilians around—is what causes you issues. When you have to leave a threat out there to cause harm or kill people. You don't get that kind of pressure release.[53]

That fight-or-flight arousal seems counterintuitive because the Reaper crews themselves are not in physical danger. However, it is prompted because they also know that in many situations, if they are not absolutely precise with their shots, non-combatants or allied forces on the ground, or both, will die. Compounding that pressure is the awareness that multiple audiences are watching in real time, from the Operations Room on the squadron to the Combined Air Operations Centre in Qatar, to other intelligence agencies who are gathering (or providing) information. Vet 15 describes being very aware—during lethal weapon events—of the current and subsequent audience:

> lots of people are watching what's going on. You can't hide things because they are visible to people who have the access to the [live video feed of the incident] you are dealing with.[54]

Experimental psychologists have studied—and continue to study—human performance under pressure. Beilock and Carr offer an insight into how 'audience awareness' might affect performance:

> Distraction-based accounts of suboptimal performance propose that performance pressure shifts attentional focus to task-irrelevant cues—such as worries about the situation and its consequences. In essence, this shift of focus changes what was single-task performance into a dual-task situation.[55]

Knowing that, under the well-intended squadron and RAF culture of learning from one another's mistakes (or successes), the video of your strike will be shown publicly in the morning briefing adds significant pressure to an already adrenaline-fuelled situation for some operators. Repeated high adrenaline spikes during the course of a shift, combined with pressure from continuous, intense external scrutiny, is one possible factor in the extreme fatigue some Reaper personnel experience. Grossman describes the 'parasympathetic backlash' that soldiers experience after the fight-or-flight response activates and spikes of adrenaline are released:

---

53   Lee, *Reaper Force*, 111.
54   Interviewee Vet 15 is a veteran of the Reaper Force.
55   Sian L Beilock and Thomas H Carr, 'On the Fragility of Skilled Performance: What Governs Choking under Pressure', *Journal of Experimental Psychology* 130, no. 4 (2001): 714.

The parasympathetic backlash occurs as soon as the danger and the excitement is over, and it takes the form of an incredibly powerful weariness and sleepiness on the part of the soldier.[56]

This qualitative research indicates a variation of response across Reaper personnel when it comes to both adrenaline response and parasympathetic backlash, which is—in turn—influenced by training, experience, psychological conditioning over time, and individual physiological and psychological traits. Consequently, the next question for future research asks: *How can physiological and psychological states be developed to maximise human performance in remote operations?*

Visual intimacy and its consequences

While the pilots or crews of other aircraft types like the Typhoon or Tornado will experience the same physiological responses to weapon use as Reaper crews, further related areas for future research are prompted in remotely piloted operations. These include the greater degree of image acuity and sustained visual intimacy, the potential for visual trauma and the extent to which visual and mental proximity to targets influences the dynamic of killing. In addition, as well as describing the physical, extreme adrenal response to a weapon strike, Ross (see above) also introduced another important factor in individual and collective Reaper culture and identity— that of 'protector' or 'guardian'.[57] Self-identification with a protective role can be in either the immediate and specific sense of direct intervention as a self-defence action (i.e. defence of non-combatants or 'friendlies' on the ground), or in a more general sense of protecting societies in Syria/Iraq from the ideology and behaviour of ISIS and its fighters. The protective role has a strong ethical basis and is a powerful motivator for Reaper personnel. It also opens up the possibility of either psychological harm or moral injury if individuals feel that they have failed to protect those that they feel responsible for.

---

56 Grossman, *On Killing*, 69. There is not the scope in this chapter for a full study of this phenomenon.
57 At a research feedback session with 13 Squadron on 14 November 2017, in response to my observation of the importance of the 'protector' role for Reaper and its personnel, a Royal Marine attached to the Squadron suggested 'Guardian' as his alternative, preferred way of conceptualising the Reaper role.

A 2014 King's College London study of mental health in the UK Armed Forces found 'no evidence of a tidal wave of deployment related mental health problems'.[58] There was no specific focus on Reaper operators in that study, which found:

> There is no evidence that the length of a single tour, or number of tours, has had an adverse effect on Service personnel's mental health, provided that Harmony Guidelines are followed. When the actual tour length exceeds the expected length, it has a substantial adverse impact on mental health and also alcohol misuse.[59]

Experiences of individuals with multiple, continuous Reaper operational tours may prompt new findings in a follow-up study. However, with PTSD rates for personnel deployed on the ground in Afghanistan between 2007 and 2009 reported at 6.9 per cent, anecdotal evidence from my Reaper research interviews does not suggest much variation from that level. This, in turn, does not—yet, anyway—diverge from levels of PTSD in civilian populations, with Norris and Slone concluding:

> It is clear that only a fraction of people who are exposed to trauma develop the full syndrome of PTSD. Thus, despite the high prevalence of trauma exposure around the world, the lifetime prevalence of PTSD is no more than 7 per cent.[60]

On the Reaper Force, as with any diverse population, a range of responses to traumatic incidents has emerged. At one end of the spectrum, a small number of cases of PTSD have been reported. Towards that end of the spectrum, Toby, a former sensor operator, presented himself at the mental health unit of a Ministry of Defence Hospital Unit. He has a traumatic memory that affects him but:

> I don't have enough ticks in the boxes for it to be full PTSD ... I still have the odd dream where I wake up with the 'bang' of a particular explosion.[61]

58 King's Centre for Military Health Research and Academic Department of Military Mental Health, *The Mental Health of the UK Armed Forces* (October 2014), 2, web.archive.org/web/20220403110441/ http://www.kcl.ac.uk/kcmhr/publications/Reports/Files/mentalhealthsummary.pdf.

59 King's Centre for Military Health Research and Academic Department of Military Mental Health, *The Mental Health of the UK Armed Forces*, 2.

60 Fran H Norris and Laurie B Slone, 'Understanding Research on the Epidemiology of Trauma and PTSD', *PTSD Research Quarterly* 24, no. 2–3 (2013): 4.

61 Lee, *Reaper Force*, 296–297.

Rory, another sensor operator, encapsulates the mixed feelings of many crew members, who live with extreme tiredness, regularly witness horrific events and yet see these as a 'normal' part of life on the Reaper Force. He observes, of himself and others:

> If anybody on the Reaper fleet says it doesn't affect them, then they're lying. It does. It has to. But I really enjoy what I do and I don't think I would change anything about the last five years with the Reaper. Five years is more than enough in one go. I am well aware that I am probably six to twelve months overdue a rest—just physically and psychologically.[62]

Many others have echoed that sentiment: there are effects, but these are bearable for most of the people, most of the time. However, over time the number of traumatic, or potentially traumatic, events accumulate: sights that cannot be unseen, incidents that stick in the mind. To add to the complexity of individual experiences on the Reaper Force, there are a small number of those who are apparently unaffected, or just minimally affected, by their experiences. Ken is no longer flying the Reaper but the influence of his experience persists:

> If I ever watch a TV programme such as Traffic Cop with camera footage from a helicopter, I immediately feel my heart rate start to increase and my mind will start to plan how best to conduct a weapon engagement on whatever person or vehicle is being tracked. Having spent several years looking at the world from, quite literally, a different angle, that perspective is, evidently, ingrained within my psyche and without hesitation is a skill-set that I am subconsciously eager to continue to employ. In all honesty I perversely miss the satisfaction and thrill of thinking on my feet, then planning and conducting a successful engagement; I'm content to say I actually enjoyed it.[63]

In a separate interview, Ken's wife reinforced what he had said: that he seemed relatively unaffected, given his experiences, and that he had enjoyed his time on 39 Squadron. Only a very small number of former Reaper personnel were as adamant as this about their enjoyment levels and the lack of impact upon them.

---

62   Lee, *Reaper Force*, 299.
63   Lee, *Reaper Force*, 294.

One of the difficulties, then, in making sense of individual reactions across the Reaper squadrons is that there is no common pattern of behaviour or response. It would appear that those with a fast jet background—who had been trained to use, or actually used, weapons—were generally well equipped to cope with the reality of killing and viewing traumatic incidents. They will have had the advantage of a lengthy fast jet training over several years where, from the outset, they would know that shooting and killing would be part of their future. Inculcation into Harrier or Tornado culture and operations would include that life and death dimension, which would not be found in the same way on maritime patrol, transport or most helicopter fleets (except Apache). However, and contradicting that generalisation, others have come from non-kinetic backgrounds—and without the years of psychological conditioning that comes with it—and excelled on the Reaper, both in technical proficiency and in psychological endurance. Trying to make sense of this inconsistency prompts another question for future research: *To what extent can individuals be socialised and conditioned to conduct lethal operations or observe traumatic events from Reaper, and to what extent do inherent levels of empathy shape individual responses?* From all of the interviews and collective engagement with Reaper personnel and spouses/ partners, I would suggest that individuals' inherent capacity for empathy is part of the mental wellbeing equation.

Empathy and lethal operations: Costs and benefits

The vast majority of people experience empathy: 'Affective or emotional empathy is when you feel along with the other person.'[64] McGregor summarises the work of Baron-Cohen in both the Empathy Bell Curve (Figure 8.2), and his 6-point Empathy Spectrum (detailed below).[65]

---

64   Simon Baron-Cohen, *The Science of Evil: On Empathy and the Origins of Cruelty* (Philadelphia: Basic Books, 2012), cited in Jane McGregor, 'The Highly Empathetic', *The Society for Research into Empathy, Cruelty & Sociopathy* (blog), May 2018, web.archive.org/web/20190905110655/http://www. sorecs.org/blog/.
65   McGregor, 'The Highly Empathetic'.

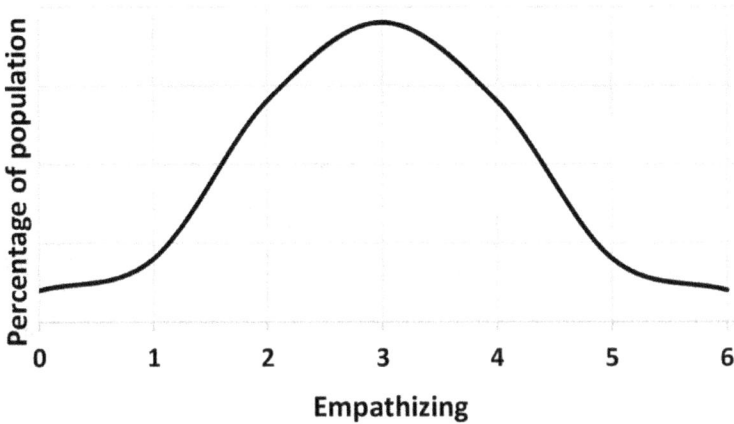

Figure 8.2: Empathy Bell Curve.
Source: Derived by Jane McGregor from Simon Baron-Cohen.

Empathy Spectrum:

Point 0   No empathy and hurting others means nothing to them

Point 1   Capable of hurting other people but feels some regret if they do so

Point 2   Has enough empathy to inhibit them from acts of physical aggression

Point 3   Compensates for lack of empathy by covering it up

Point 4   Low to average empathy

Point 5   Slightly higher than average empathy

Point 6   Very focused on the feelings of others. An almost unstoppable drive to empathise.[66]

As the Empathy Bell Curve and Empathy Spectrum suggest, there is typically a small number of people in a population who experience an overwhelming degree of empathy with others (Point 6 on the spectrum). Two interviewees shared how they had each, independently, reached a point where they could no longer kill a human target, even when that person was positively identified as an enemy fighter, a legal target, and the strike was correctly authorised. Jake described his thought process in choosing which of two people to strike, in a shot that his crew was legally authorised to take but which could not hit both targets with the same missile or bomb:

---

66   McGregor, 'The Highly Empathetic'.

> What right do I have to decide which one of you is going to live for another sixty years, have children and grandchildren, and which one of you is going to have your life ended right now? Do I have the right to make that decision?[67]

His language and tone indicated a strong capacity for empathy—putting himself in their places—which influenced his thinking and actions.

Meanwhile, at Point 0 or 1 on the spectrum, there are a small number who react the opposite way, with little or no empathy (or the capacity to somehow disengage it), and who are able to kill with little or no personal approbation. Consider the self-description from a special forces soldier:

> In a fight, physical or verbal, it felt like I was detached. It was almost like I was watching myself in slow motion and thinking clearly about what needed to be done and how I was going to do it. There was no fear, no emotional connection to what was happening.[68]

On two occasions, almost identical descriptions of being 'in the zone' before and during a Reaper weapon strike have been shared with me. In February 2018, one very experienced crew member[69] described being 'absolutely cold' and his heart rate 'barely moving' when striking a human target (while also confirming that he complied with ROE). Was that Reaper crew member congenitally predisposed and lacking empathy, or conditioned over time and through experience? Or a combination of those factors?

The previous two examples appear to be rare, with the overwhelming majority having a capacity for empathy somewhere between the extremes that not only allowed them to conduct Reaper operations but also helped them to do so. Empathy can play a number of roles in day-to-day squadron life and the conduct of operations. Empathy can be a powerful motivator, especially in a protective role, in providing personal and professional fulfilment on the Reaper Force. It can also lead to powerful self-questioning if that protective instinct is violated. For Jamie, a mission intelligence coordinator, protecting allies on the ground in Afghanistan was a strong motivator, rooted in his own prior experience out on the ground. His first weapon event was against a Taliban vehicle that was transporting explosives

---

67 Lee, *Reaper Force*, 165–166.
68 Kevin Dutton and Andy McNab, *The Good Psychopath's Guide to Success* (London: Bantam Press, 2014), 18.
69 Research visit to squadron. For anonymity I will not even give his crew position.

but resulted in several civilian deaths, which he observed in detail.[70] His thoughts at the time capture both his desire to make a positive difference through his work and the immediate impact upon him when his aims and expectations were violated:

> How did I find myself in this situation? I joined the Reaper Force to make a positive difference after the shit I experienced on the ground in Afghanistan. How did my first weapon event turn into a nightmare, an awful nightmare?[71]

As well as watching the initial impact, Jamie and the crew also continued observing the area for hours afterwards and watched the bodies being removed from the destroyed vehicle.

More research will be needed to more fully understand the extent and role of empathy in enabling or limiting individual ability to conduct lethal, remote air operations. Lawrence et al. set out to measure empathy, with some success. However, they caution that 'it is important to tease out the different kinds of emotional reactivity and distinguish between empathic and other types of emotional responses'.[72] Separately, Head explores the costs of empathy in the international political arena with the aim of demonstrating 'how it is frequently costly for those who make the ethical-political choice to engage in empathy in situations of conflict and political violence'.[73] This cost will vary according to the degree of empathy of the individual concerned and, potentially, the degree to which an incident also reflects or violates personal ethics. Every situation has the potential for great complexity. For example, a Reaper pilot might have a high capacity for empathy and be strongly affected by witnessing a public beheading that (s)he was unable to disrupt. As well as the potential for visually mediated mental trauma (note that trauma is not assured), core ethical principles may be violated, thereby inducing a dissonant state. It is such a violation of personal ethics or core beliefs that prompt the final consideration in this chapter: the possibility of moral injury among Reaper personnel.

---

70   The 2011 civilian casualty (CIVCAS) incident was acknowledged by the RAF and Ministry of Defence. It is described in full from Jamie's perspective in the chapter 'CIVCAS' in *Reaper Force*.
71   Lee, *Reaper Force*, 107.
72   Emma J Lawrence, P Shaw, D Baker, Simon Baron-Cohen, and Anthony S David, 'Measuring Empathy: Reliability and Validity of the Empathy Quotient', *Psychological Medicine*, no. 34 (2004): 919.
73   Naomi Head, 'Costly Encounters of the Empathic Kind: A Typology', *International Theory* 8, no. 1 (2016): 176.

Moral injury

The term 'moral injury' (Nash et al.'s definition having been set out earlier in the chapter) has gained a significant public and academic profile in the 21st century. Edgar Jones points out that 'there is no agreed definition of moral injury', which makes the subject difficult to explain and explore from the outset.[74] In an extensive and growing literature, Drescher et al. suggest: 'The term that has been used to describe the impact of various acts of omission or commission in war that produces inner conflict is *moral injury*.'[75] Meanwhile, Frankfurt et al. refer to it as:

> a transdiagnostic syndrome that describes the uniquely deleterious impact of committing or failing to prevent acts during warfare that involve violations, transgressions, or betrayals, of commonly accepted boundaries of behavior.[76]

All of these definitions presuppose a moral dimension to war, usually with each side claiming moral superiority over the other,[77] and each morality offers the possibility of violation.

The moral context for RAF Reaper operations is bounded by *jus in bello* in the operational domain and by the UK government's *jus ad bellum* justification for deploying air power against anti-government forces such as the Taliban in Afghanistan, ISIS in Iraq and Syria, and against the Syrian government in response to the use of chemical weapons.[78] While air strikes have had some public support, it has not been overwhelming. For example, in December 2015 Prime Minister David Cameron proposed extending attacks against IS from Iraq to Syria.[79] While 48 per cent of respondents in one poll supported the action, with 30 per cent opposed, there was not

---

74   Edgar Jones, 'Moral Injury in Time of War', *The Lancet* 391 (5 May 2018): 1767, doi.org/10.1016/S0140-6736(18)30946-2.

75   Kent D Drescher et al., 'An Exploration of the Viability and Usefulness of the Construct of Moral Injury in War Veterans', *Traumatology* 17, no. 1 (2011): 8.

76   Sheila B Frankfurt, Patricia Frazier, and Brian Engdahl, 'Indirect Relations between Transgressive Acts and General Combat Exposure and Moral Injury', *Military Medicine* 182 (Nov/ Dec 2017): e1950.

77   An example of contested moral superiority can be found in Tom Smith, Peter Lee, Vladimir Rauta, and Sameera Khalfey, 'Understanding the Syria Babel: Moral Perspectives on the Syrian Conflict from Just War to *Jihad*', *Studies in Conflict & Terrorism* 43, no. 12 (2020): 1108–1128, doi.org/10.1080/1057610X.2018.1523358.

78   Ethical challenges faced by RAF Reaper operations are explored in Peter Lee, 'Rights, Wrongs and Drones: Remote Warfare, Ethics and the Challenge of Just War Reasoning', *Air Power Review* 16, no. 3 (Autumn/Winter 2013): 30–49.

79   IS: Islamic State, previously called ISIS.

a clear majority in favour.[80] Moral injury becomes possible, according to Drescher, Frankfurt and others if social norms and the moral framework of a combatant are violated, either by an action, or lack of action, on their part. It also becomes possible if they witness acts that breach personal morality, such as executions, rape or murder, that they cannot prevent.

Numerous variables are associated with moral injury: PTSD symptoms, self-injury, demoralisation and self-handicapping.[81] Further, and more obviously linked to the term 'moral injury', Maguen and Litz suggest 'an act of serious transgression that leads to serious inner conflict because the experience is at odds with core ethical and moral beliefs'.[82] Other factors that are linked to moral injury include shame, guilt, transgression of spiritual or religious beliefs, self-condemnation, social problems, trust issues and spiritual/existential issues.[83] This very brief summary highlights the possibility that the concept of moral injury is in danger of being extended so far as to be too vague to be of practical or explanatory use. However, for the purpose of meaningful application to members of the Reaper Force, I suggest that 'moral injury' is most likely to be applicable in the following situations: powerless witnessing (unable to intervene to stop a heinous act), observations that grievously violate the watcher's social norms and core beliefs (e.g. beheadings) and unintended ethical transgression (such as unintentional harming of a non-combatant). Consider three examples that could *potentially* contribute to moral injury, so defined.[84]

First, where individuals have witnessed in close detail the deaths of, or physical harm to, allied combatants or non-combatants for whom they feel a protective responsibility, but were powerless to intervene because of legal constraints (ROE), potential secondary threats to yet other allies or non-combatants, or because they could not see or anticipate the threat to life.

> That flash on the screen [as their vehicle exploded], and the feeling of impotence, just stayed with us. Our job was to provide overwatch on these guys, to protect them. We had been staring, looking for

80   Ben Glaze, 'David Cameron Fails to Convince Public to Back Strikes against ISIS in Syria', *Daily Mirror*, 2 December 2018, www.mirror.co.uk/news/uk-news/david-cameron-fails-convince-public-6914446?ICID=FB_mirror_main.

81   Brett T Litz et al., 'Moral Injury and Moral Repair in War Veterans: A Preliminary Model and Intervention Strategy', *Clinical Psychology Review* 29, no. 8(2009): 695–706.

82   Shira Maguen and Brett Litz, 'Moral Injury in Veterans of War', *PTSD Research Quarterly* 23, no. 1 (2012): 1.

83   Maguen and Litz, 'Moral Injury', 1–2.

84   To reiterate, these examples are illustrative and not definitive. There are too many complicating factors that would have to be excluded for moral injury to be confirmed.

anything that might be a threat. But there was a big puddle over the junction, and the soldiers with their hand-held detectors couldn't see [the explosive device] underneath. That whole incident has stuck with me. I'm not saying I have PTSD, but I'm saying that I get how some people are affected that way.[85] (Johnny, Pilot)

Second, the violation of social norms. Sensor operator Jake struggled with the idea of taking life, but also struggles with the fact that he eventually adapted to it. Despite being able to make the logical, intellectual calculation of the relative moral benefit in killing jihadists who would otherwise inflict harm on others, the taking of life left a deep impression on him:

> I hate the fact that I've killed people. I hate the fact that I've taken life in a very calculating manner. But I also hate the fact that I seem to be able to live with it now.[86]

Then, third, there is unintended ethical transgression—even where all of the legal and operational checks and approvals have been correctly observed. Jamie, mentioned above, and his crew had acted on the best available intelligence when they destroyed the Taliban vehicle full of explosives, yet still had to live with the unintended civilian deaths:

> I wanted to continue because for me it was always for some greater good. And degrading the enemy's capability was the primary role. Certainly not to injure or kill anybody that wasn't the enemy, and certainly not women and children, that's for sure. I never knew how I'd feel if that would happen, and again it's difficult to explain. You never forget. You never want to forget.[87]

Much work continues to be done on 'moral injury', and it is offered here as a potential explanation of some responses by Reaper personnel to experiences while conducting remote air operations. This chapter argues that the immersive visual, emotional and psychological aspects of Reaper operations play a part in individualised responses to surveillance activities and weapon events in recent counterinsurgency actions. In relation to the foregoing sections and arguments, two factors relating to moral injury are suggested here for further future study. They emerged during data gathering by the author with several members of the Reaper Force and assume that individualised capacity for empathy is an important consideration. First,

---

85   Lee, *Reaper Force*, 75–76.
86   Lee, *Reaper Force*, 172.
87   Lee, *Reaper Force*, 111–112.

that moral injury may be caused to more empathetic members of the Force by the way that less empathetic individuals speak about the enemy and killing. And, second, for highly empathetic individuals (see the Empathy Bell Curve), ethics education is important in order for them to make sense of their actions in an operational-ethical context. For less- or non-empathetic individuals, ethics education may be even more essential, so that they can appreciate—intellectually if not emotionally—how they can negatively impact their colleagues. Such research may well provide insights and understanding that translate to other situations where visually traumatic events are encountered, from military imagery analysis to civilian website content moderators.

Conclusion

There have been enormous technological advances in the delivery of air power since the First World War days of the Royal Flying Corps, the Royal Naval Air Service and the advent of the RAF. Despite the technical developments that enable the Reaper to be operated across continents, war and air operations remain essentially human activities. This chapter has focused on that human dimension, identifying emotional, psychological and moral complexities that emerge from the distance–intimacy paradox of remote air warfare. Yet even though much has changed, much remains the same. In March 1917, James McCudden reflected, in a language and tone that is now often parodied: 'sometimes one sits and thinks, "Oh, this damned war and its cursed tragedies".'[88] Yet a year later and not long before his own death he shot down an enemy German aircraft and recorded: 'As I looked at the machine I saw the enemy gunner fall away from the Hannover fuselage. I had no feeling for him for I knew he was dead.'[89] Even in the culturally coded language of the early 20th century, the change in character and mental condition over time can be detected.

The close-up views afforded to Reaper personnel as they conduct ISR and strike operations make the act of taking human life more visually intimate and therefore more psychologically and emotionally involved than it has ever been before for aircrew in the history of the RAF. It is important not to assume that psychological harms are either inevitable or long-lasting

---

88   McCudden, *Flying Fury*, 130.
89   McCudden, *Flying Fury*, 239.

for everyone involved: they are not. However, it is equally important to recognise that a new dynamic has been introduced to the delivery of air power that can negatively impact remote operators—some more than others, and for reasons we do not yet fully understand. As the RAF enters its second century, the distance paradox brought about by the advent of the Reaper should not be underestimated, especially if remotely piloted aircraft are set to outnumber manned counterparts in the coming years. The potential jeopardy of manned air operations is replaced by increased psychological and emotional jeopardy in remotely piloted air operations. This could be dangerous in an organisation that has, throughout its history, deliberately downplayed emotional and psychological risks and reactions almost to the point of parody. *Per ardua ad astra*, states the motto of the RAF: 'through adversity to the stars'. For RAF remote aircrew in the next hundred years, adversity—and achievement—will take place on the ground. Supporting and developing those remote operators is the next great challenge.

References

Bandura, Albert. 'Disengaging Morality from Robotic War'. *The Psychologist* 30 (February 2017): 39.

Bandura, Albert. 'Moral Disengagement in the Perpetration of Inhumanities'. *Personality and Social Psychology Review* 3, no. 3 (1999): 193–209. doi.org/10.1207/s15327957pspr0303_3.

Baron-Cohen, Simon. *The Science of Evil: On Empathy and the Origins of Cruelty*. Philadelphia: Basic Books, 2012.

BBC. 'Syria War: MoD Admits Civilian Died in RAF Strike on Islamic State'. *BBC*, 2 May 2018, www.bbc.co.uk/news/uk-43977394.

Beilock, Sian L, and Thomas H Carr. 'On the Fragility of Skilled Performance: What Governs Choking under Pressure'. *Journal of Experimental Psychology* 130, no. 4 (2001): 701–725. doi.org/10.1037//0096-3445.130.4.701.

Benjamin, Medea. *Drone Warfare: Killing by Remote Control*. New York and London: OR Books, 2012.

Blair, Dave, and Karen House. 'Avengers in Wrath: Moral Agency and Trauma Prevention for Remote Warriors'. *Lawfare*, 12 November 2017, www.lawfareblog.com/avengers-wrath-moral-agency-and-trauma-prevention-remote-warriors.

Cavallaro, James , Stephan Sonnenberg, and Sarah Knuckey. *Living under Drones: Death, Injury and Trauma to Civilians from US Drone Practices in Pakistan.* September 2012, International Human Rights and Conflict Resolution Clinic (Stanford Law School) and the Global Justice Clinic (NYU School of Law), law.stanford.edu/publications/living-under-drones-death-injury-and-trauma-to-civilians-from-us-drone-practices-in-pakistan/.

Cheshire, Leonard. *Bomber Pilot.* London: HarperCollins, 1975.

Cole, Chris, Mary Dobbing, and Amy Hailwood. *Convenient Killing: Armed Drones and the 'Playstation' Mentality.* Oxford: Fellowship of Reconciliation, 2010.

Drescher, Kent D, David W Foy, Caroline Kelly, Anna Leshner, Kerrie Schutz, and Brett Litz. 'An Exploration of the Viability and Usefulness of the Construct of Moral Injury in War Veterans'. *Traumatology* 17, no. 1 (2011): 8–13. doi.org/10.1177/1534765610395615.

Dutton, Kevin, and Andy McNab. *The Good Psychopath's Guide to Success.* London: Bantam Press, 2014.

Farmer, Ben. 'RAF Reaper Drone Interrupts Islamic State Public Execution'. *The Telegraph,* 16 May 2017, www.telegraph.co.uk/news/2017/05/16/raf-reaper-drone-interrupts-islamic-state-public-execution/.

Frankfurt, Sheila B, Patricia Frazier, and Brian Engdahl. 'Indirect Relations between Transgressive Acts and General Combat Exposure and Moral Injury'. *Military Medicine* 182 (Nov/ Dec 2017): e1950–e1956. doi.org/10.7205/MILMED-D-17-00062.

Garrett, Stephen A. *Ethics and Airpower in World War II: The British Bombing of German Cities.* New York: St. Martin's Press, 1993.

Glaze, Ben. 'David Cameron Fails to Convince Public to Back Strikes against ISIS in Syria'. *Daily Mirror,* 2 December 2018, www.mirror.co.uk/news/uk-news/david-cameron-fails-convince-public-6914446?ICID=FB_mirror_main.

Grossman, Dave. *On Killing: The Psychological Cost of Learning to Kill in War and Society.* Rev. ed. New York: Back Bay Books, 2009.

Hays Parks, W. '"Precision" and "Area" Bombing: Who Did Which, and When?' *Journal of Strategic Studies* 18, no. 1 (1995): 145–174. doi.org/10.1080/01402399508437582.

HC 772 Defence Committee. 'Written Evidence from the Royal Aeronautical Society'. *parliament.uk*, 15 September 2013, publications.parliament.uk/pa/cm201314/cmselect/cmdfence/772/772vw13.htm.

Head, Naomi. 'Costly Encounters of the Empathic Kind: A Typology'. *International Theory* 8, no. 1 (2016): 171–199. doi.org/10.1017/S1752971915000238.

House of Commons Defence Committee. *Remote Control: Remotely Piloted Air Systems—Current and Future UK Use: Government Response to the Committee's Tenth Report of Session 2013–14*. Sixth Special Report of Session 2014–15. London: House of Commons, 20 July 2014, publications.parliament.uk/pa/cm201415/cmselect/cmdfence/611/611.pdf.

Jones, Edgar. 'Moral Injury in Time of War'. *The Lancet* 391 (5 May 2018): 1766–1767. doi.org/10.1016/S0140-6736(18)30946-2.

Kelman, Herbert G. 'Violence without Moral Restraint: Reflections on the Dehumanization of Victims and Victimizers'. *Journal of Social Issues* 29, no. 4 (Fall 1973): 25–61. doi.org/10.1111/j.1540-4560.1973.tb00102.x.

Killeen, Damian, and David Jordan. 'RPAS: Future Force or Force Multiplier? An Analysis of Manned/Unmanned Platforms and Force Balancing'. *Air Power Review* 16, no. 3 (2013): 22–23.

King's Centre for Military Health Research and Academic Department of Military Mental Health. *The Mental Health of the UK Armed Forces*. October 2014, web.archive.org/web/20220403110441/http://www.kcl.ac.uk/kcmhr/publications/Reports/Files/mentalhealthsummary.pdf.

Lawrence, Emma J, P Shaw, D Baker, Simon Baron-Cohen, and Anthony S David. 'Measuring Empathy: Reliability and Validity of the Empathy Quotient'. *Psychological Medicine*, no. 34 (2004): 919.

Lee, Peter. *Reaper Force: Inside Britain's Drone Wars*. London: John Blake Publishing, 2018.

Lee, Peter. 'Return from the Wilderness: An Assessment of Arthur Harris's Moral Responsibility for the German City Bombings'. *Air Power Review* 16, no. 1 (Spring 2013): 70–90.

Lee, Peter. 'Rights, Wrongs and Drones: Remote Warfare, Ethics and the Challenge of Just War Reasoning'. *Air Power Review* 16, no. 3 (Autumn/Winter 2013): 30–49.

Lee, Peter, and Colin McHattie. 'Churchill and the Bombing of German Cities 1940–1945'. *Global War Studies* 13, no. 1 (2016): 47–69. doi.org/10.5893/19498489.130102.

Litz, Brett T, Nathan Stein, Eileen Delaney, Leslie Lebowitz, William P Nash, Caroline Silva, and Shira Maguen. 'Moral Injury and Moral Repair in War Veterans: A Preliminary Model and Intervention Strategy'. *Clinical Psychology Review* 29, no. 8 (2009): 695–706. doi.org/10.1016/j.cpr.2009.07.003.

Maguen, Shira, and Brett Litz. 'Moral Injury in Veterans of War'. *PTSD Research Quarterly* 23, no. 1 (2012): 1.

McCammon, Sarah. 'The Warfare May be Remote but the Trauma Is Real'. *NPR*, 24 April 2017, www.npr.org/2017/04/24/525413427/for-drone-pilots-warfare-may-be-remote-but-the-trauma-is-real.

McCudden, James B. *Flying Fury*. Poland: Amazon Fulfilment, 2009.

McGregor, Jane. 'The Highly Empathetic'. *The Society for Research into Empathy, Cruelty & Sociopathy* (blog), May 2018, web.archive.org/web/20190905110655/http://www.sorecs.org/blog/.

Ministry of Defence. 'New Investment in Counter Terrorism for UK Armed Forces'. *GOV.UK*, 4 October 2015, www.gov.uk/government/news/new-investment-in-counter-terrorism-for-uk-armed-forces.

Nash, William P, Jennifer Vasterling, Linda Ewing-Cobbs, Sarah Horn, Thomas Gaskin, John Golden, William T Riley, Stephen V Bowles, James Favret, Patricia Lester, Robert Koffman, Laura C Farnsworth, and Dewleen G Baker. 'Consensus Recommendations for Common Data Elements for Operational Stress Research and Surveillance: Report of a Federal Interagency Working Group'. *Archives of Physical Medicine and Rehabilitation* 91, no. 11 (2010): 1673–1683. doi.org/10.1016/j.apmr.2010.06.035.

Norris, Fran H, and Laurie B Slone. 'Understanding Research on the Epidemiology of Trauma and PTSD'. *PTSD Research Quarterly* 24, no. 2–3 (2013): 4.

Olsthoorn, Peter. *Military Ethics and Virtues: An Interdisciplinary Approach for the 21 Century*. New York: Routledge, 2011. doi.org/10.4324/9780203840825.

Power, Matthew. 'Confessions of a Drone Warrior'. *GQ*, 22 October 2013, www.gq.com/story/drone-uav-pilot-assassination?currentPage=4.

Press, Eyal. 'The Wounds of the Drone Warrior'. *New York Times*, 13 June 2018, www.nytimes.com/2018/06/13/magazine/veterans-ptsd-drone-warrior-wounds.html.

Royal Air Force. 'Reaper (MQ-9A)'. *Royal Air Force*, www.raf.mod.uk/aircraft/reaper-mq9a/, accessed 15 June 2018.

Singer, Peter W. *Wired for War*. New York: Penguin, 2009.

Smith, Tom, Peter Lee, Vladimir Rauta, and Sameera Khalfey. 'Understanding the Syria Babel: Moral Perspectives on the Syrian Conflict from Just War to Jihad'. *Studies in Conflict & Terrorism* 43, no. 12 (2020): 1108–1128, doi.org/10.1080/1057610X.2018.1523358.

Volpato, Chaiara, and Alberta Contarello. 'Towards a Social Psychology of Extreme Situations: Primo Levi's *If This Is a Man* and Social Identity Theory'. *European Journal of Social Psychology* 29 (1999): 239–258. doi.org/10.1002/(SICI)1099-0992(199903/05)29:2/3<239::AID-EJSP926>3.0.CO;2-O.

Webster, Charles, and Noble Frankland. *The Strategic Air Offensive against Germany 1939–1945 Vol. IV.* London: Her Majesty's Stationery Office, 1961.

Wright, Gerald. 'The Storm Shadow Cruise Missile'. *UK Defence Journal*, 20 October 2015, ukdefencejournal.org.uk/the-storm-shadow-cruise-missile/.

9

The Role of an Operational Frame in Furthering the International Debate on Lethal Autonomous Weapons Systems

Ian MacLeod and Erin Hahn

Introduction[1]

For over five years, the international community has been engaged in a debate about the potential regulation of lethal autonomous weapon systems (LAWS). The United Nations Group of Governmental Experts on LAWS continues to be the main forum for the international dialogue, which remains splintered about the need (or not) for regulation.[2] While the focus of the debate has shifted over time, the fundamental point of contention is really whether the development and fielding of LAWS removes the human role in the decision to kill, and if so, whether the consequences (operational, moral, legal) are acceptable.

---

1   Editors' note: as mentioned in the Introduction, this chapter was written in 2020, and global events since then may have overtaken some aspects. As the pace of technological and cultural change has continued to accelerate, the editors have opted to present this text as drafted to minimise further delays in publication.
2   United Nations, 'Background on Lethal Autonomous Weapons Systems in the CCW', *United Nations*, web.archive.org/web/20200820214625/https://www.unog.ch/80256EE600585943/(httpPages)/8FA3C2562A60FF81C1257CE600393DF6?OpenDocument, accessed 22 February 2020.

The debate over LAWS is not correctly framed to analyse lethal decisions or the role that machines play in those decisions, which is a major reason why the forum is not generating fruitful debate on the key issue at hand. To address this framing deficiency, our work introduced an operational lens through which to evaluate existing weapons systems and example cases of LAWS, specifically using the dynamic targeting cycle. Modern militaries using advanced weapon systems distribute decision-making over multiple entities across time, space, echelon and authority. Framing the issue of LAWS as making the decision, or as if there is only a single, penultimate decision to kill, represents an incomplete understanding of how lethal decisions are made. Decisions about warfare start as national priorities and then cascade down as derivative decisions in the form of operational strategies, tactical planning and ultimately tactical execution. Additionally, the language and terminology routinely used in arguments about LAWS, and artificial intelligence (AI) more generally, anthropomorphise the technology in a way that conveys a superficial understanding of technical capabilities and promotes flawed analogies. Is the machine making a decision, or is it reacting or generating a response based on input data in accordance with its programmed goals? Many have gravitated toward the former, as sophisticated systems can give the appearance of agency. The international debate has failed to expose the dangers of this perception due to the lack of an operational and technical understanding of how the weapons will work and be fielded, which is the critical foundation on which effective policy frameworks can be built.

Decision-making

In modern conflicts between states, decisions over life and death are the end result of a long sequence of derivative decisions that start at the national level with the choice to go to war. The goals and strategy of the proposed military operation emerge from that national decision as well as how that country thinks about conflict from a doctrinal perspective. The goals and strategy of the operation then drive and bind the targeting decisions that will be made and the rules of engagement that will be followed. Individual missions and the attacks that occur within them all flow from these prior choices. The terms 'selection and engagement' of targets has been a focus of the LAWS debate, with the implication that the entirety of the lethal decision occurs in a single moment. But those terms are really tactical ones within a sortie or mission, and are a refinement of all the many prior decisions that have prioritised the type of targets to be struck, where and when those strikes should occur, battlefield coordination measures and the rules of engagement to be followed.

Modern state militaries have developed elaborate processes to plan and execute complex operations, and these processes are the means by which these prior decisions are incorporated and followed to ensure that desired operational outcomes are achieved within the limits of national values. Ekelhof has described the deliberate targeting process followed by the North Atlantic Treaty Organization (NATO) and select other countries in detail, and the International Panel for the Regulation of Autonomous Weapons (iPRAW)[3] has demonstrated how the complementary dynamic targeting process could apply to LAWS in the future.[4] The key feature of these processes is that lethal decisions are distributed and the end result is conditional on prior choices. Processes have evolved and been modified over time as new technology has been brought to bear on military problems, but the fundamental distribution of decisions has endured. Newer technologies have simply created new choices and options based on precision over longer ranges, tracking and guidance capabilities and other advancements. Decision-making processes and control measures have adjusted to enable new capabilities to meet the intent of the decision-makers.

Ethical camps

In the course of our participation with iPRAW, we found the use of the targeting cycle to evaluate potential LAWS very helpful in evaluating the potential for autonomy in weapon systems. It also led to a more structured discussion around the legal and ethical issues with specific applications of LAWS. Many in the group agreed it was possible LAWS could in some circumstances be used in compliance with international humanitarian law, and that there were some obvious boundaries, given the state of technology that presented legal problems. However, over time it became clear that

---

3   For more about iPRAW, see the iPRAW homepage: web.archive.org/web/20220313064416/ https://www.ipraw.org/; iPRAW, *Focus on the Ethical Implications for a Regulation of LAWS*, Focus on Report no. 4 (International Panel on the Regulation of Autonomous Weapons, August 2018).
4   Merel AC Ekelhof, 'Lifting the Fog of Targeting: "Autonomous Weapons" and Human Control through the Lens of Military Targeting', *Naval War College Review* 71, no. 3 (2018): Article 6; iPRAW, *Focus on the Human–Machine Relation in LAWS*, Focus on Report no. 3 (International Panel on the Regulation of Autonomous Weapons, March 2018).

the bigger disagreement was not with lawful use as much as the ethical implications of LAWS. Much of the disagreement centres on the concept of human dignity and how it is or is not violated by LAWS.

We identified three ethical camps that were influencing the debate: scepticism, consequentialist and deontological. The scepticism camp does not believe human dignity exists or if it does, it is not relevant to the debate subsumed as a part of other human rights. For some, LAWS should be evaluated strictly through international humanitarian law absent a tie to the concept of dignity or to broader human rights law. It is argued that LAWS are machines and not moral agents and therefore cannot make the moral judgement to take human life. In contrast, the consequentialist camp does not necessarily deny that LAWS could violate human dignity, but it allows the benefits of using LAWS to supersede any such violation if, on balance, the benefits outweigh the negative consequences. This is also referred to as utilitarianism, which is a prominent consequentialist theory. Last, the deontologist camp is focused on the morality of human choices. Human rights, including dignity, are inherent and their violation cannot be justified by beneficial outcomes.

On the point of dignity explicitly, the term is referenced in numerous international treaties and state documents (e.g. the German Constitution). While nothing in international humanitarian law refers specifically to the violation of dignity, many consider it implied. Common criteria for dignity derive from the Martens Clause, which requires that the entity taking the life has the ability to recognise a human being as such, and with the attendant rights that come with being human; has an understanding of the value of life and the significance of its loss; and can reflect on the reasons for taking life and determine it is justified.[5] In accordance with this definition, for deontologists, autonomy in weapon systems breaks the link to moral agency and violates human dignity. For consequentialists, if they ascribe to this definition, a violation can be overridden by a larger good. And for sceptics, it does not matter. While the group achieved an intellectual understanding of these differences, it was not until we applied it to a specific case through a thought experiment that the key boundaries became clear.

---

5   See Peter Asaro, 'Jus Nascendi: Robotic Weapons and the Martens Clause', in *Robot Law*, ed. Ryan Calo, A. Michael Froomkin, and Ian Kerr (Edward Elgar Publishing, 2016), 367–386, doi.org/10.4337/9781783476732.00024.

Thought experiment[6]

We used a thought experiment with the diverse group of iPRAW researchers on the LAWS topic. The experiment helped bring clarity to the philosophical underpinnings of the disagreement over LAWS and decision-making in conflict. It involved a simple ground combat (mortarman) scenario with successive iterations that added more technology to the main weapon in use. The purpose of the thought experiment is to identify the point at which a philosophical split emerges on how autonomous function in a weapon is viewed. The group walked through each iteration and then asked if the description violated an ethical tenet or principle of international humanitarian law (IHL). Figure 9.1 illustrates these three iterations.

Figure 9.1: Mortarman thought experiment scenarios.
Source: Drawn by the authors.

Imagine a ground combat scenario between two conventional forces in a declared conflict. A mortar crew could be ordered to direct fire against an enemy mechanised infantry unit located over a ridgeline based on coordinates and directions relayed from another unit with knowledge of the target and target area. The mission calls for destroying an armoured personnel carrier that the other unit has reported with an approximate location. In this baseline case, the crew could fire a series of mortar rounds in a small pattern around a coordinate at the approximate location because its sister unit is both under threat and has reported that the area is clear of non-combatants. The intent of the firing would be to destroy the armoured personnel carrier. In this case, the mortar round itself is controlled by the crew's adherence to firing procedures only. The group agreed that this scenario would be legal under IHL and ethical, even though the mortar would not be aimed precisely at a specific target, but rather the coordinate where the target was estimated to be located.

---

6   This thought experiment was developed by Deane-Peter Baker and used as part of the iPRAW Working Group in development of iPRAW, *Focus on the Ethical Implications for a Regulation of LAWS*.

The next iteration added a global positioning system (GPS) feature to the mortar round that ensured the round would go to the intended coordinate even if the round was blown by wind or some other external stimulus interfered with the mortar crew's intent. This too was deemed both legal and ethical. The addition of this feature was not viewed as decision-making by a machine but enhanced precision that could improve the ability of the weapon system to achieve the intended result.

The last iteration added a visual sensor to the mortar round that would employ machine learning-enabled image recognition to identify the desired target, an armoured personnel carrier, and correct the mortar's impact point onto the target. If the automated target recognition did not sense one of the specified types of adversary targets, it would self-destruct. This feature revealed the point of divide among group members. The additional sensor was viewed as delegating the decision to kill to a machine, and not as an additional means of meeting the intent of the mortar crew.

A striking point about this last scenario is that those who felt the target recognition feature broke the link to moral agency (deontological camp) accepted that if the technology worked correctly, it could be more precise and cause less collateral damage. Nonetheless, they would prefer the first scenario with less precision and potentially more casualties than the violation of dignity they perceive in the last scenario. The sceptics were clearly in favour of the third scenario and the consequentialists were on board if it led to a more beneficial outcome than the first two scenarios. Additionally, the technical function described in the last scenario differs only a little from many terminal guidance seekers that have been used for decades such as heat-seeking or radar-guided missiles that home in on and adjust their course in this terminal phase. The deontological camp makes a distinction between this type of technology and the notional applications of AI into weapon systems to accomplish similar goals even though the practical function is the same in both cases—to more precisely destroy the intended target when direct control is not possible. This view makes finding common ground challenging.

Reframing the debate

We argue that framing LAWS as machines making decisions about life and death is misplaced based on a misunderstanding of technology and how modern warfare is carried out. While hypothetical LAWS are likely to have profound impacts to how future warfare is conducted, they will have to comport with how states carry out the complex decision-making involved with targeting. Thinking about LAWS as capabilities that enable new options for military commanders to carry out their intent, or put another way, to extend control into situations where they previously had none, refocuses the discussion onto the real dilemmas and challenges that will need to be faced when deciding on development and use of LAWS.

The ethical camps of consequentialists and sceptics see a path forward that builds off command and control doctrine and practices as the means by which to incorporate LAWS into combat operations in ways that are operationally effective and compliant with IHL. Further, the consequentialist camp sees the potential for LAWS to enable precision in a way that could lead to less collateral damage. Both camps fundamentally see LAWS, or rather the technological capabilities that it would represent, as means by which the intent of the commander can be extended. However, those paths rest on the ability of technologists to develop systems that can be predictable and reliable enough to enable commanders to make informed decisions about the type of missions, environments and targets that LAWS are appropriate to be employed against. Other technological advances that have been incorporated into controlling weapons such as GPS guidance and active seekers, to name but a couple, all have empirically based performance characteristics to help make those informed judgements, and the procedures for planning and executing combat operations with them has evolved to account for both the advantage and risks they incur. The deontologists see LAWS in a fundamentally different way and it appears no amount of technical progress or operational effectiveness will change that view. They see LAWS not as an extension of human intent, but a delegation or interruption in the human decision-making process surrounding life and death.

These differences exist and will persist. The value of applying an operational lens through which to analyse the appropriateness of potential LAWS is that it lays bare the points of departure and narrows the discussion to the dilemmas that deserve thoughtful debate. Deontologists' concern about breaking the link to moral agency is an important question consequentialists and sceptics need to address whether they agree with it on its face or not. That point was most clearly articulated through an operational viewpoint—via the analysis of the mortarman example and command and control doctrine. Likewise, the practical benefits LAWS could bring to the battlefield (e.g. fewer civilian casualties) is something deontologists need to grapple with, and an analysis of current technology as compared to future LAWS helps identify where there are truly differences and where the similarities may weigh in favour of at least limited use of LAWS. Again, the focus on what is operationally desirable and technically feasible forces all camps onto practical ground and helps avoid debates in the abstract over concepts and ideas that modern militaries are ever more urgently being forced to confront. This kind of analysis is designed not to point to the answers, but to clearly articulate the key issues and trade-offs with LAWS so states can make policy decisions based on a thorough and informed debate.

References

Asaro, Peter. 'Jus Nascendi: Robotic Weapons and the Martens Clause'. In *Robot Law*, edited by Ryan Calo, A Michael Froomkin, and Ian Kerr, eds, 367–386. Edward Elgar Publishing, 2016. doi.org/10.4337/9781783476732.00024.

Ekelhof, Merel AC. 'Lifting the Fog of Targeting: "Autonomous Weapons" and Human Control through the Lens of Military Targeting'. *Naval War College Review* 71, no. 3 (2018): Article 6.

iPRAW. *Focus on the Ethical Implications for a Regulation of LAWS*. Focus on Report no. 4. International Panel on the Regulation of Autonomous Weapons, August 2018.

iPRAW. *Focus on the Human–Machine Relation in LAWS*. Focus on Report no. 3. International Panel on the Regulation of Autonomous Weapons, March 2018.

United Nations. 'Background on Lethal Autonomous Weapons Systems in the CCW'. *United Nations*, web.archive.org/web/20200820214625/https://www. unog.ch/80256EE600585943/(httpPages)/8FA3C2562A60FF81C1257 CE600393DF6?OpenDocument, accessed 22 February 2020.

10

Lessons from Tradition: Just War, Wisdom and Restraint in a Changing World

Valerie Morkevičius

Introduction[1]

The classical just war tradition finds itself under attack from all sides these days. On the one hand, the rise of a populist version of *realpolitik* denies the very relevance of ethical limits on warfighting. Consider, for example, President Trump's vocal hostility to the laws of war, his administration's loosening of the regulations on the use of drones, and its interest in pardoning several American military personnel accused or convicted of war crimes. From this point of view, just war thinking—and international law for that matter—is nothing more than a 'politically correct' impediment to victory.

On the other hand, classical just war thinking finds itself in the crosshairs of those who should be its fellow travellers—particularly revisionist just war thinkers, who claim that the tradition lacks both analytical and ethical rigour.[2] The heart of the revisionist argument is that traditional

---

1   Editors' note: as mentioned in the Introduction, this chapter was written in 2020, and global events since then may have overtaken some aspects. As the pace of technological and cultural change has continued to accelerate, the editors have opted to present this text as drafted to minimise further delays in publication.

2   Jeff McMahan, 'The Sources and Status of Just War Principles', *Journal of Military Ethics* 6, no. 2 (2007): 91–106, doi.org/10.1080/15027570701381963.

ways of thinking about just war fail to appropriately consider the rights of individuals.[3] The pragmatic and communitarian impulses of the classical tradition are seen as leading to morally untenable conclusions. From this perspective, classical just war thinking is worse than useless—it's morally bankrupt.

In short, the classical tradition finds itself at an ironic crossroads, decried simultaneously as too idealistic and too pragmatic to be of use. At the same time, changes in the way modern wars are fought have left strategists and ethicists alike with the unsettled feeling that our existing frameworks may not be adequate to face future challenges ranging from hybrid wars to artificial intelligence (AI). And yet, the world's three great just war traditions—Christianity, Islam and Hinduism—have between them some 3,600 years of combined experience thinking about the ethics of war in a constantly evolving world. The persistence of these traditions in the face of multiple revolutions in military affairs suggests they warrant a second look. In a world where the future of warfare looks to be so different from what we've known—where cyberwarfare and information warfare speed up time and erase state borders, and where space warfare and autonomous weapons challenge the limits of humanity itself—going back to the past is a useful way to look again into the future. Drawing broadly on these three traditions, I argue that there are at least seven important lessons we can draw from the past. Thus, while the historical just war traditions are imperfect, we should be careful to avoid throwing out the baby with the bathwater.

There was no golden age

Critics of the just war tradition—and of ethical and legal approaches to limiting the harms of war more generally—sometimes suggest that our existing ethical and legal frameworks are no longer applicable in a world in which the line between 'war' and 'not war' is blurred, and in which new technologies and tactics further blur the line between combatants and non-combatants. Pundits and academics alike are fond of decrying our era as uniquely fraught, as scarier and more morally complex than any that came

---

3   Helen Frowe, *Defensive Killing* (Oxford: Oxford University Press, 2014), doi.org/10.1093/acprof:oso/9780199609857.001.0001; Jeff McMahan, 'Proportionality and Necessity in *Jus in Bello*', in *The Oxford Handbook of Ethics of War*, ed. Seth Lazar and Helen Frowe (Oxford: Oxford University Press, 2016), doi.org/10.1093/oxfordhb/9780199943418.013.24.

before. We seem to be surrounded by revolutions in military affairs that threaten to upend all of our prior knowledge and assumptions: cyberwarfare, information warfare, hybrid warfare, space warfare and even AI.

Furthermore, these new revolutionary capabilities themselves are subject to debate, and definitions differ as to what they cover, adding to confusion. For instance, hybrid war is a contested term, and while the concept might not be truly new, it wasn't until recently that scholars attempted to give it precision. Frank Hoffman in 2007, examining the Second Lebanese War of 2006, defined hybrid war as 'the incorporation of a range of different modes of warfare, including conventional capabilities, irregular tactics and formations, terrorist acts, including indiscriminate violence and coercion and criminal disorder'.[4] This type of conflict fuses 'the lethality of state conflict with the fanatical and protracted fervor of irregular warfare'. The term covers the manner of conflict and the manner of their organisation, both of which may use conventional and non-conventional aspects.[5] Asymmetric tactics are nothing new, but increasingly this form of warfare is seen to challenge state actors and their responses, and it tests the boundaries of ethical and legal approaches.

Revolutionary claims are also made for cyberwar. However, here too clarity is elusive. While used extensively in public discourse, a commonly agreed definition has yet to be found.[6] Indeed, analysis diverges significantly, suggesting on the one hand it is a direct, imminent threat while on the other arguing that it is not 'war' in any meaningful way.[7] Ashraf codifies these as alarmist, sceptic and realist, and concludes that a broadly agreed-upon scholarly definition of cyberwar is unlikely to emerge.[8] Robinson et al. suggest that cyberwar and cyberwarfare are poorly differentiated, while the

---

4   Frank Hoffman, *Conflict in the 21st Century: The Rise of Hybrid Wars* (Arlington: Potomac Institute for Policy Studies. 2007), 8.

5   Hoffman, *Conflict in the 21st Century,* 28.

6   See Michael Robinson, Kevin Jones, and Helge Janicke, 'Cyber Warfare: Issues and Challenges', *Computers & Security* 49 (2015): 70–94, doi.org/10.1016/j.cose.2014.11.007; Cameran Ashraf, 'Defining Cyberwar: Towards a Definitional Framework', *Defense & Security Analysis* 37, no. 3 (2021): 274–294, doi.org/10.1080/14751798.2021.1959141.

7   As an example of the former, see Richard A Clarke and Robert Knake, *CyberWar: The Next Threat to National Security and What to Do about It*, reprint ed. (New York: Ecco, 2012). For the latter, see Thomas Rid, *Cyber War Will Not Take Place* (Oxford and New York: Oxford University Press, 2013).

8   Ashraf, 'Defining Cyberwar', 286. Wide variations are noted in considerations of the actions, the actors, the effects, the geography and the targets.

cyber weapons themselves elude precise definition, being difficult to identify and presenting difficulties in establishing how they may be controlled or limited, and certainly complex in their ethical implications.

But claims that war has been rendered forever different because of some new technology or tactic isn't new—it's as old as the crossbow, if not older. Every generation's attempt to gain the advantage by designing weapons that can be targeted from a greater distance, with more accuracy and ever greater destructive power—from the crossbow to rifles to radios to tanks to airplanes to missiles to nuclear weapons to drones to cyber tools and AI systems— has been seen (at least by somebody) as a revolution in military affairs that has forever changed the existing way of war, altering the predicted pattern of winners and losers, and ultimately shifting the global balance of power. And while the pace of change may be faster today, we should take comfort in knowing that the changes of the past were pretty disconcerting for those who lived through them too.

Surprisingly, the historical just war traditions have had little to say about the permissibility of particular weapons. While military strategists and legal scholars often react with panicked excitement to technological change— think about the Second Lateran Council banning crossbows in 1139— scholars working within the just war tradition have been virtually silent. In fact, it's hard to even find references to specific weaponry in any of the canonical texts. Of all the Christian just war thinkers, only one—Francisco de Vitoria in 1557—mentions a specific weapon type, and then only in passing.[9] A handful of Islamic scholars, including al-Shafi'i, al-Misri and al-Mawardi, refer to mangonels.[10] And Hindu texts describe tactics, but gloss over the details of the weapons used. What accounts for this relative silence? Perhaps these long-ago scholars recognised something that is hard for us to see in our tech-obsessed world: the real moral problems lie in the ways new technologies are used—and not in the technology itself.

9   Francisco de Vitoria, *Vitoria: Political Writings* (Cambridge: Cambridge University Press, 1991), 316, doi.org/10.1017/CBO9780511840944.
10   Khaled Abou El Fadl, *Rebellion and Violence* (Cambridge: Cambridge University Press, 2003), 152; Ahmad ibn Naqib al-Misri, *Reliance of the Traveler: A Classic Manual of Islamic Sacred Law,* trans. Nuh Hah Mim Keller (Beltsville: Amana Publications, 1994), 594; Al-Mawardi, *Adab al-dunya wa al-din,* ed. Mustafa al-Saqqa (Cairo: Mustafa Babi al-Halabi and Sons, 1955), 79.

There won't be a world without war

A serious just war thinker should keep company with plenty of pacifists and just peace thinkers. The danger of a tradition, such as just war thinking, that believes that some wars can be justified because some wars can do more harm than good is that it's easy to be lulled into the assumption that one's own wars always meet the standard.[11] Those who question the very possibility that war can ever be justified, that it can ever accomplish any good, can force us to confront our own blind spots. This sort of moral pacifism can be found in every major religious tradition, and hinges on the faith that wrongs on Earth will be met by a divine response—and that it is not the place of human beings to judge each other or to use force to punish such wrongdoings themselves. There is an honest and faithful humility in such pacifism, and just war thinkers have much to learn from such approaches.

But just war thinkers themselves are at risk of subscribing to an institutional, back-door pacifism that can be equally dangerous. This isn't truly moral pacifism, but a sort of pragmatic pacifism. It allows that force could be legitimate, but only if carried out by ideal, non-biased judges. In this view, war is a tool that should be reserved for ideal agents—perhaps the United Nations, or at the very least, a coalition of states who are somehow able to rise above their inherently rooted and complicated national identities.[12] War, in this view, has been legislated to the edge of existence by international law—a sort of extreme lawsuit that can only be decided by actors who somehow stand above the fray of politics. Yet the international system lacks just the sort of robust legal institutional structure that would be necessary to make such a worldview make sense.[13] And the inherent anarchy of the world system makes the construction of such a structure vanishingly unlikely.

Fundamentally, this institutionalist view contradicts the realism at the heart of just war thinking. Instead, just war thinking recognises war as an unfortunate, persistent reality in a world of competitive states and bad guys

---

11   John Howard Yoder, *When War Is Unjust: Being Honest in Just-War Thinking* (Eugene: Wipf and Stock Publishers, 2001); Andrew Fiala, *The Just War Myth: The Moral Illusions of War* (Maryland: Rowman & Littlefield, 2008).

12   George R Lucas Jr, 'The Role of the "International Community" in Just War Tradition—Confronting the Challenges of Humanitarian Intervention and Preemptive War', *Journal of Military Ethics* 2, no. 2 (2003): 134, doi.org/10.1080/15027570310000261; Fiala, *The Just War Myth*, 12.

13   John J Mearsheimer, 'The False Promise of International Institutions', *International Security* 19, no. 3 (1994): 5–49, doi.org/10.2307/2539078.

of all stripes.[14] In the Christian tradition, the fallen nature of mankind makes the use of force permissible—even necessary. In Augustinian terms, human nature is such that 'all men desire to be at peace with their own people, while wishing to impose their will upon those people's lives'.[15] While Islamic scholars believe that Allah forgave Adam and Eve, their account of human nature remains pessimistic, as pride makes it difficult for humans to submit to Allah's will, and thus to live in peace with each other. Likewise, the Hindu tradition warns us that it is only the imposition of law and punishment by princes that prevents humans from behaving like beasts.[16] Given humankind's conflictual nature, and our species' unfortunate penchant for greed, pride and hate, we cannot expect to see permanent peace in our time.

There is no such thing as a war that doesn't hurt the wrong people

Since the late 19th century, the laws of war surrounding the treatment of individuals—civilians, medical personnel and prisoners of war—have grown denser and more elaborate. At the same time, our broader understanding of human rights has also grown, raising many important questions about *in bello* just war ethics.[17] In essence, these changes reflect a shift in the way we think about the ethics of war, moving the just war debate away from its theological roots towards more philosophical and legal approaches, a change which has been afoot since the 18th century. These changes have brought about immense improvements in the lives of individuals around the world in peacetime and in war, and at all points in between.

Some contemporary just war thinkers, especially those of the revisionist school, have adopted this individualist framework. Consequently, their work identifies troubling aspects of war that might have been overlooked by scholars in the past. If force can be used against another human being because of some grave wrong that has been committed, what do we do

---

14   Valerie Morkevičius, *Realist Ethics: Just War Traditions as Power Politics* (Cambridge: Cambridge University Press, 2018), 26, doi.org/10.1017/9781108235396.

15   Augustine of Hippo, *City of God* (New York: Penguin Books, 1984), 867.

16   Maganlal A Buch, *The Principles of Hindu Ethics* (Baroda: Arya Sudharak Printing Press, 1921), 353.

17   Cecile Fabre, 'Cosmopolitanism, Just War Theory and Legitimate Authority', *International Affairs* 84, no. 5 (2008): 963–976, doi.org/10.1111/j.1468-2346.2008.00749.x; Mary Kaldor, 'From Just War to Just Peace', in *The Viability of Human Security*, ed. Monica den Boer and Jaap de Vilde (Amsterdam: Amsterdam University Press, 2008), 36.

with the knowledge that not every armed combatant on the other side is there willingly? Some have been drafted, some have been brutally coerced, some may be children.[18] Are such individuals liable to be harmed? And are they really *more* deserving of harm than certain civilians who may be eager supporters of their regime's policies, and whose work may—especially in this cyber era—be even more relevant to the war effort?[19] Or what do we do with the knowledge that in fighting against a brutal and repressive regime, or against warlords with no interest in human rights at all, we will inevitably end up harming some civilians who had nothing at all to do with the conflict at hand, and who might even be victims of those inhumane leaders themselves?

These are not simple questions, and their answers are not easy or comfortable. Some revisionists might have us believe that given the moral risk fighting in war poses, we had better not fight any wars at all. This sort of response to the complex reality of war is a form of human rights pacifism.[20] It's laudable, in that it reminds us of the seriousness of war—it is true that no war is ever clean. No war will ever be fought without harming or killing at least some individuals who did nothing to deserve it. And that fact should give us pause. It should encourage us to adopt an attitude of restraint, an awareness that even in trying to do good, we will cause pain. It should also encourage us to treat warfighting with gravitas: war and killing aren't glorious, or fun, or funny.

But while we must always be deeply, painfully aware of the grave responsibility of wielding force, we mustn't let this awareness paralyse us. To say that all war is wrong because all war harms the innocent is to allow evil to proceed uninhibited. It would be cold comfort to say to those facing genocide or serious crimes against humanity or cruel occupation that we can't use force to defend them because, in so doing, we will hurt other innocent people. And the communal life we build at home, too, is the sort of good that deserves to be protected. Using force for these good ends will have unintended negative consequences—but these sorts of good ends may be worth those losses. Choosing to use force is morally challenging, but we shouldn't allow that to cow us into a pacifism that arises not from moral convictions, but from fear of messing up.

---

18 Jeff McMahan, *Killing in War* (Oxford: Oxford University Press, 2009), 33–35, doi.org/10.1093/acprof:oso/9780199548668.001.0001.
19 Cécile Fabre, *Cosmopolitan War* (Oxford: Oxford University Press, 2012), doi.org/10.1093/acprof:oso/9780199567164.001.0001.
20 Morkevičius, *Realist Ethics*, 216.

Death isn't the worst of war

Contemporary just war arguments often frame the ethics of war in terms of rights and corresponding duties. States have a right to use force in certain circumstances; states also have a duty to protect their citizen soldiers. Civilians have a right to be protected; soldiers have a duty to protect them. This perspective sees the wrongs of war as moments in which someone has infringed on someone else's rights. We do wrong in war when we treat others in ways they aren't liable to be treated. And this is certainly the way both international law and revisionist just war thinking frame the problem of war.

But the earlier just war traditions understood the moral wrongs that happen in war not only as a matter of what we do to others but also as a matter of what war does to us. As Augustine put it, for a people who believe in the promise of a life to come, the problem of war can't be 'that those who will die someday are killed so that those who will conquer might dominate in peace'.[21] The real tragedy of war lies instead in 'the desire for harming, the cruelty of revenge, the restless and implacable mind … the lust for dominating', and other inhumane sentiments that war brings out in us.[22]

This matters in particular as we consider a future of warfare in which we hand over certain aspects of decision-making to AI systems, including, possibly, the use of lethal autonomous weapons. What will outsourcing lethal decisions do to us as moral actors? Will entrusting AI systems with the power to make life-and-death decisions affect the way we value human life more broadly? Will it ultimately make us see others—or even ourselves— as more disposable? Being killed by a person or by a code makes little difference to the person who dies—but what does that knowledge do to his or her comrades, family and community? What will postwar reconciliation look like after a conflict in which the killing has been done by robots?

The historical just war tradition helps us recognise that, as we consider new technologies and tactics, we must ask not only whether they can be used according to existing ethical principles and international law—in other words, what the effects of our use of such technologies and tactics will be *on others*—but also what the effects of their use will be on those of us who use them, as moral creatures. Even when killing from a distance, human beings

---

21   Augustine of Hippo, *Political Writings* (Indianapolis: Hackett, 1994), 221.
22   Augustine of Hippo, *Political Writings*, 221.

remain conscious of their actions. The sombre responsibility of killing, even from a distance, weighs heavily on many soldiers, as evidenced by the high rates of post-traumatic stress disorder (PTSD) and moral injury among drone operators, for example.[23]

Ethics isn't maths: There isn't a single right answer

Individuals who are optimistic about the ethical future of warfare in a world in which more and more aspects of warfighting are handed over to automated systems, or even AI, often argue that such algorithms can be programmed with the relevant legal and ethical codes. Once such systems 'know the rules', they will be even better than humans at ethical decision-making, as their judgement will never be clouded by fatigue, fear, anger or hatred.[24]

The mistake in such reasoning is the assumption that moral reasoning about war can (and should) be reducible to a set of programmable principles. First, when we speak about the ethics of war—or even the just war tradition—we are never really speaking in the singular. Multiple systems of ethics and multiple approaches to the just war tradition exist, even 'just' in the Western context. The tradition's lack of systematisation is the natural product of its organic growth over several millennia, as it blended the logics of several traditions: Roman law, chivalric warrior ethics, and the work of the early Church fathers and medieval Scholastics. Today's legalist and revisionist just war thinkers add the logic of law and continental philosophy to the mix.

That complexity only grows when we consider non-Western approaches. While these diverse approaches share similar questions and basic intuitions about the need to limit the scope of war, they differ, for example, in the way they define which individuals may be targeted, and which may not, in their approach to weighing necessity and proportionality and, more broadly, in what constitutes a just cause.

23   Christian Enemark, 'Drones, Risk, and Moral Injury', *Critical Military Studies* 5, no. 2 (2019): 150–167, doi.org/10.1080/23337486.2017.1384979; Peter Lee, *Reaper Force: Inside Britain's Drone Wars* (London: John Blake Publishing, 2018).

24   Ronald Arkin, *Governing Lethal Behavior in Autonomous Robots* (New York: Taylor and Francis); Patrick Lin, 'Ethical Blowback from Emerging Technologies', *Journal of Military Ethics* 9, no. 4 (2010): 313–331, 313, doi.org/10.1080/15027570.2010.536401.

Second, this diversity both within and between traditions has inherent value. When confronted with other ethical approaches that differ from our own, we are forced to examine our own underlying assumptions. We grow into mature ethical thinkers who can examine specific dilemmas from multiple perspectives—in Technicolor 3D if you will—rather than a more simplistic black-and-white 2D approach.

In other words, this lack of systematisation is a feature, not a bug. Consider, for example, the problem of what philosopher Michael Walzer terms 'awkward combatants'. In *Just and Unjust Wars*, Walzer describes the moral unease felt by First World War combatants when, upon leaping into the enemy's trenches, they came across enemy combatants in the midst of ordinary activities, like shaving or relieving themselves.[25] Another 'awkward combatant' is the soldier who is armed, but frozen by fear. What are we to do with such individuals?

While individuals working within the classical Western just war tradition may intuit—as the soldiers in Walzer's examples do—that something moral is at stake, their approach lacks the vocabulary to adjudicate these types of situations. It's not a problem 'covered' by the usual canon. And yet the concept is not foreign to the West: the logic for avoiding killing armed but temporarily helpless combatants can be found in the chivalric literature, embedded in the concept of warrior's honour. Likewise, the revisionist tradition's approach to violence through a rights-based lens can help us think seriously about how the real threat posed by a particular individual might vary by time and context.

The study of other ethical traditions can also help us recognise new-to-us ethical conundrums, and to develop a richer vocabulary for analysing them. For example, the Hindu tradition distinguishes between objective and subjective threats. The *Mahabharata* thus enjoins readers:

> In this world, the slaughter of sleeping persons is not applauded … The same is the case with persons that have laid down their arms and come down from cars and steeds. They also are unslayable who say 'We are thine!' and they that surrender themselves, and they whose locks are dishevelled, and they whose animals have been killed under them or whose cars have been broken.[26]

25 Michael Walzer, *Just and Unjust Wars* (New York: Basic Books 1977), 143–144.
26 *Mahabharata of Krishna-Dwaipayana Vyasa*, Book 10, trans. Kisari Mohari (BiblioLife, 2009).

While unprepared soldiers—those 'whose locks are dishevelled' or whose means of transport have been destroyed—are objectively threats, they are not subjective threats (or at least not to the same degree) as they would be if they were capable of their full lethality. While attacking them might be necessary, the *Mahabharata*'s reminder that killing such individuals isn't praiseworthy should give us pause. Are there alternate ways of achieving our tactical goals, without killing them? Would killing them undermine our broader strategic aims? Would their deaths at our hands leave us questioning whether we had done the right thing? Can we live with those questions, or will they haunt us long after the fighting is done?

None of these questions are easy to answer, and, arguably, none of these questions has a *definitive* answer. Given that war and ethics are quintessentially human activities, that shouldn't surprise us. We shouldn't be afraid to argue and to disagree—such debates are a sign that we are taking our moral responsibilities seriously. While humans are capable of shifting between multiple ethical frameworks and moving between levels of analysis with relative ease, AI is unlikely to be able to do the same. For this reason, while we may be willing to embrace AI systems to improve our ability to identify and precisely attack targets, we should be hesitant about embracing any system that claims to be able to do our ethical work for us.

Ethical action requires character

Because the right action in war isn't reducible to a simple set of rules, fighting ethically requires character. While Augustine and Aquinas are often referred to as the founding fathers of the Christian just war tradition, they actually didn't have much to say about the right behaviour in war. Aquinas warns us not to kill innocents and leaves it at that. So what else—besides rules alone—is necessary for generating ethical action? According to the classical just war traditions of Christianity, Islam and Hinduism, just action flows naturally from just actors. Virtue produces virtuous behaviour.

But how do we constitute ourselves as the sorts of actors who will do as much right as possible? Virtue requires 'the development of good inclinations— we are virtuous when doing the right thing gives us pleasure'.[27] Indeed, the US military in particular has adopted a 'rule-bounded virtue ethics'

---

27   Peter Olsthoorn, *Military Ethics and Virtues: An Interdisciplinary Approach for the 21st Century* (London: Routledge, 2010), 5, doi.org/10.4324/9780203840825.

approach in professional military education.[28] To be effective, however, virtue education cannot involve mere classroom exercises and reading assignments. As McIntyre puts it, in a rather tongue-in-cheek way:

> we do not become virtuous by attending lectures on ethical theory or, worse still, going to conferences on ethics, but by being educated into the relevant habits of thought and action in the life of everyday practice.[29]

Conversations about the right action must be a part of daily life.

One shortcoming of ethics education within contemporary professional military education is that while it does (in the US at least) use the language of virtue ethics, it does not embrace the concept of virtue in the broader sense that the classical just war traditions would have it.[30] Within the logic of the traditions, soldiers' virtues were not unique to their roles as soldiers—or at least, the virtues directly related to soldiering were only *some* of the virtues a soldier was supposed to possess. Virtues relevant to the military—such as courage and prudence—should not be separated from the broader set of virtues that make for good human beings, including love, temperance and justice. This suggests that conversations about character and virtue within the military must extend beyond questions directly applicable to military service to broader questions of what it means to be a good person and to live a good life.

We are all embedded in a time and place

Finally, exploring the historical just war traditions can help us recognise our own parochialism, reminding us that our own situatedness limits our moral imaginations. Each tradition of just war shapes not only our sense of the moral limits of war but also defines what even counts as an ethical problem.

Furthermore, an examination of the history of just war thinking reveals that it has always been in a conversation with power. Those of us who value the tradition the most highly sometimes imagine that this means our

28 Marcus Schulzke, *Pursuing Moral Warfare: Ethics in American, British, and Israeli Counterinsurgency* (Washington, DC: Georgetown University Press, 2019), doi.org/10.2307/j.ctvb1htwk.
29 Alasdair C McIntyre, 'Military Ethics: A Discipline in Crisis', in *Routledge Handbook of Military Ethics,* ed. George Lucas (Routledge, 2015), 4.
30 Matthew Beard, 'Virtuous Soldiers: A Role for the Liberal Arts?', *Journal of Military Ethics* 13, no. 3 (2014): 274–294, doi.org/10.1080/15027570.2014.977546.

predecessors in the tradition boldly spoke truth to power. And it's true—just war thinkers across the ages have pushed back against the dehumanisation of our enemies, against the use of civilians as pawns in war, against the comforting presumption that God is always on our side.

But there's a reason why just war thinking has persisted through the ages. Yes, just war thinking raises uncomfortable questions. But it also proposes practical answers, answers that have been useful for decision-makers precisely because they haven't demanded too much. The norms of the just war tradition reflect the interests of the powerful.[31] Consequently, the rules have changed within each tradition as different groups came to dominate politically. Both Christianity and Islam began as religions of groups who held no power—and embraced pacifism as a response. Yet when Christian and Muslim groups came to power, they each found ways to justify the kinds of wars their states needed to fight. *In bello* limitations on targeting across all three traditions likewise reflect pragmatic concerns: the limits of technology, military necessity and so on.

Knowing that the just war tradition is itself implicated in power politics doesn't mean, however, that we should dismiss it as hopelessly tainted. But it does mean that we should be cautious to avoid moral certainty. We must do our best to make moral decisions, whether how to fight, or about going to war in the first place. But we must do so with humility, knowing that future generations may deem our ethical frameworks to be lacking.

And while the temptation to presume we're standing on the moral high ground is omnipresent, it's important to remember that throughout the development of the just war traditions, thinkers have always cautioned against assuming that we are truly able to perceive what justice means in a particular case. We must do our best. But humans are fallible, information is often incomplete and decision-making can be clouded by a variety of psychological pressures. The just war traditions thus remind us that we must never give up on asking tough questions of ourselves, even returning to questions we think we've already answered.

The wars of the future—whether we want to imagine them as hybrid wars, or cyberwars, or information wars—will certainly differ from the wars our ancestors fought. Nonetheless, the critical ethical questions remain eternal. For this reason, there is still a great deal to be learned from dusting off

---

31 Morkevičius, *Realist Ethics*.

those old tomes of traditional just war thinking. When we stand on the shoulders of those who came before, we can see a little farther—and from a different angle—than we might have otherwise. Even when we find the arguments of those who came before to be wanting, even when we disagree vehemently with their conclusions, these debates across eras and cultures can challenge us to better justify our own moral presuppositions. For those of us interested in just war thinking, this suggests that while the classical traditions may no longer be fashionable, they may still have some important lessons left to impart.

References

Abou El Fadl, Khaled. *Rebellion and Violence*. Cambridge: Cambridge University Press, 2003.

Al-Mawardi. *Adab al-dunya wa al-din*. Edited by Mustafa al-Saqqa. Cairo: Mustafa Babi al-Halabi and Sons, 1955.

Al-Misri, Ahmad ibn Naqib. *Reliance of the Traveler: A Classic Manual of Islamic Sacred Law*. Translated by Nuh Hah Mim Keller. Beltsville: Amana Publications, 1994.

Arkin, Ronald. *Governing Lethal Behavior in Autonomous Robots*. New York: Taylor and Francis.

Ashraf, Cameran. 'Defining Cyberwar: Towards a Definitional Framework'. *Defense & Security Analysis* 37, no. 3 (2021): 274–294. doi.org/10.1080/14751798.2021.1959141.

Augustine of Hippo. *City of God*. New York: Penguin Books, 1984.

Augustine of Hippo. *Political Writings*. Indianapolis: Hackett, 1994.

Beard, Matthew. 'Virtuous Soldiers: A Role for the Liberal Arts?' *Journal of Military Ethics* 13, no. 3 (2014): 274–294. doi.org/10.1080/15027570.2014.977546.

Buch, Maganlal A. *The Principles of Hindu Ethics*. Baroda: Arya Sudharak Printing Press, 1921.

Clarke, Richard A, and Robert Knake. *CyberWar: The Next Threat to National Security and What to Do about It*. Reprint ed. New York: Ecco, 2012.

de Vitoria, Francisco. *Vitoria: Political Writings*. Cambridge: Cambridge University Press, 1991. doi.org/10.1017/CBO9780511840944.

Enemark, Christian. 'Drones, Risk, and Moral Injury'. *Critical Military Studies* 5, no. 2 (2019): 150–167. doi.org/10.1080/23337486.2017.1384979.

Fabre, Cécile. 'Cosmopolitanism, Just War Theory and Legitimate Authority'. *International Affairs* 84, no. 5 (2008): 963–976. doi.org/10.1111/j.1468-2346.2008.00749.x.

Fabre, Cécile. *Cosmopolitan War.* Oxford: Oxford University Press, 2012. doi.org/10.1093/acprof:oso/9780199567164.001.0001.

Fiala, Andrew. *The Just War Myth: The Moral Illusions of War.* Maryland: Rowman & Littlefield, 2008.

Frowe, Helen. *Defensive Killing.* Oxford: Oxford University Press, 2014. doi.org/10.1093/acprof:oso/9780199609857.001.0001.

Hoffman, Frank. *Conflict in the 21st Century: The Rise of Hybrid Wars.* Arlington: Potomac Institute for Policy Studies. 2007.

Kaldor, Mary. 'From Just War to Just Peace'. In *The Viability of Human Security*, edited by Monica den Boer and Jaap de Vilde, 21–46. Amsterdam: Amsterdam University Press, 2008.

Lee, Peter. *Reaper Force: Inside Britain's Drone Wars.* London: John Blake Publishing, 2018.

Lin, Patrick. 'Ethical Blowback from Emerging Technologies'. *Journal of Military Ethics* 9, no. 4 (2010): 313–331. doi.org/10.1080/15027570.2010.536401.

Lucas, George R, Jr. 'The Role of the "International Community" in Just War Tradition—Confronting the Challenges of Humanitarian Intervention and Preemptive War'. *Journal of Military Ethics* 2, no. 2 (2003): 122–144. doi.org/10.1080/15027570310000261.

Mahabharata of Krishna-Dwaipayana Vyasa. Book 10. Translated by Kisari Mohari. BiblioLife, 2009.

McIntyre, Alasdair C. 'Military Ethics: A Discipline in Crisis'. In *Routledge Handbook of Military Ethics*, edited by George Lucas, 3–14. Routledge, 2015.

McMahan, Jeff. *Killing in War.* Oxford: Oxford University Press, 2009. doi.org/10.1093/acprof:oso/9780199548668.001.0001.

McMahan, Jeff. 'Proportionality and Necessity in *Jus in Bello*'. In *The Oxford Handbook of Ethics of War*, edited by Seth Lazar and Helen Frowe, 418–439. Oxford: Oxford University Press, 2016. doi.org/10.1093/oxfordhb/9780199943418.013.24.

McMahan, Jeff. 'The Sources and Status of Just War Principles'. *Journal of Military Ethics* 6, no. 2 (2007): 91–106. doi.org/10.1080/15027570701381963.

Mearsheimer, John J. 'The False Promise of International Institutions'. *International Security* 19, no. 3 (1994): 5–49. doi.org/10.2307/2539078.

Morkevičius, Valerie. *Realist Ethics: Just War Traditions as Power Politics.* Cambridge: Cambridge University Press, 2018. doi.org/10.1017/9781108235396.

Olsthoorn, Peter. *Military Ethics and Virtues: An Interdisciplinary Approach for the 21 Century.* London: Routledge, 2010. doi.org/10.4324/9780203840825.

Rid, Thomas. *Cyber War Will Not Take Place.* Oxford and New York: Oxford University Press, 2013.

Robinson, Michael, Kevin Jones, and Helge Janicke. 'Cyber Warfare: Issues and Challenges'. *Computers & Security* 49 (2015): 70–94. doi.org/10.1016/j.cose.2014.11.007.

Schulzke, Marcus. *Pursuing Moral Warfare: Ethics in American, British, and Israeli Counterinsurgency.* Washington, DC: Georgetown University Press, 2019. doi.org/10.2307/j.ctvb1htwk.

Walzer, Michael. *Just and Unjust Wars.* New York: Basic Books 1977.

Yoder, John Howard. *When War Is Unjust: Being Honest in Just-War Thinking.* Eugene: Wipf and Stock Publishers, 2001.

Conclusion

Mark Hilborne

This volume has explored the impact of technology and new domains on future warfare, and while several themes run throughout the chapters, they mostly highlight the increasing complexity of the security environment and the uncertainty of future war. The sense of time and speed has been, and continues to be, compressed by developments in quantum technologies, the cyber domain, artificial intelligence (AI) and the increased capabilities of sensors and data collection, as well as new propulsion technologies such as hypersonic weapon designs. Concepts regarding the shape and extent of the battlefield are challenged by the notion of hybrid war and subthreshold tactics, as well as new domains in which competition is increasing, such as space. The virtual world of the cyber domain can connect disparate elements of a group or cause, where they can vie for advantage in the 'digital narrative' and where the clear lines between decision-makers, war-fighters and non-combatants are blurred. Further challenging the shape of the battlefield is the increased development of remote and autonomous warfare. Within these developments is the increase in the commercial sector, which is leading more and more technological advances, eclipsing government and military capabilities in many fields. This will have a direct impact on how military production is owned and managed, on the composition of military forces, and, of course, on how states and their military institutions will respond in terms of strategy.

The last 20 years have already witnessed a diversification of challenges to which states have had to adapt. A rise in terrorism and insurgency has driven a need for militaries to refamiliarise themselves with counter-strategies, responding to so-called asymmetric tactics. Non-state actors have proliferated and become much more sophisticated, learning to avoid a full confrontation with conventional militaries, and thus creating an image of impotence of Western military force, while also challenging the primacy of the state as the most relevant entity in the global environment. Alongside this there has been a shift in economic power, and with that, military power, towards Asia. Furthermore, demographic growth creates social pressures that drive and aggravate resource scarcity. Into this mix we can add rapid technological development, some of which may be beneficial, but which also holds the potential for profound change to conflict that could unfold in unpredictable ways.

Claims of war's transformation are by no means new—technological change has consistently been seen as a novel force, bestowing revolutionary capabilities. Change has been an enduring feature in war, however, and very often those changes have not resulted in the revolution that has been claimed. In 1953, Cyril Fall cautioned:

> Observers constantly describe the warfare of their own age as marking a revolutionary breach in the normal progress of methods of warfare. Their selection of their own age ought to put readers and listeners on their guard … It is a fallacy, due to ignorance of technical and tactical military history, to suppose that methods of warfare have not made continuous and, on the whole, fairly even progress.[1]

Thus, a declaration of revolutionary change must be judged prudently. Nonetheless, there is now a confluence of new technologies that combine to create the potential for fundamental transformation at many levels. This wave of technological change has been called the fourth industrial revolution (4IR), first put forward by Klaus Schwab, and is characterised by an exponential rather than a linear rate of change. These technologies may not affect the fundamental nature of war as set out by Clausewitz, but they will likely impact its character. From a military perspective, the key will be the impact on the speed of operations, and on the shape of the operational domain—the factors of time and space. The combination of these shifts will increasingly affect how states approach and engage in conflict.

Operational time and space

The concept of the 4IR sees advances building upon the digital revolution of the previous wave—underway since the middle of the last century—but distinguished by a convergence and complementarity of emerging technology

---

1   Cyril Falls, *A Hundred Years of War, 1850–1950* (London: Duckworth, 1953), 13.

domains. While the impact of current and future advances affecting warfare will be subject to competing visions, 4IR certainly encompasses the breadth of developments covered in this volume and sets a framework for what may be considered War 4.0.

Both the speed at which military operations occur, and the breadth of those operations, will be transformed by this confluence of military developments. It can be argued that operational concepts have already moved from a linear and sequential process to one that is parallel in nature, in which military power is employed simultaneously at the strategic, operational and tactical levels of war in order to paralyse the enemy's ability to respond and operate. This concept, applied in the first Gulf War, was a step further than merely compressing what would previously be a sequence of actions. By abridging the levels of war, the dimensions of time and space can be exploited.[2] The simultaneity of the application of force provides the compression of time, while the scale of the attack across the entirety of a country, and a breadth of targets, expands the space. Yet these two axes are set to be affected much more profoundly by the building wave of rapid technological change.

While the capabilities of Gulf War I—also deemed revolutionary by some observers—were themselves based on a series of advances in communications technology, exploitation of space-derived data, and enhanced sensors, the current wave goes further, incorporating technology that will fuse and overlay the physical, digital and biological spheres. More than an increase in efficiency on what has gone before, 4IR draws together advances in those digital technologies, but encompasses nanotechnology, biotechnology, new materials and advanced digital production. Its impact will not only be felt in the military domain, but also more widely in society.

A central idea is the persistent interconnectedness resulting from developments in the cyber domain and mobile internet. The power and number of connected devices, and the access to knowledge that this combination provides, generates unlimited potential.[3] For military practitioners, however, there are challenges. The cyber domain, while a product of past decades, has not been a significant feature in recent conflicts. As a result, there is little in the way of existing experience, and

---

2   Brigadier General David A Deptula, *Effects-Based Operations: Change in the Nature of Warfare* (Arlington: Aerospace Education Foundation, 2001), 5.

3   Klaus Schwab, 'The Fourth Industrial Revolution: What It Means, How to Respond', *World Economic Forum*, 14 January 2016, www.weforum.org/agenda/2016/01/the-fourth-industrial-revolution-what-it-means-and-how-to-respond/.

little upon which to build best practice. To what extent can lessons from other domains act as a template? Unlike other domains, cyber is virtual, and not tangible. It may be argued, however, that it is both a domain and a set of technologies. A cyber campaign presents a different sense of time, generating a sense of immediacy, compressing the temporal element of operations and decision-making.

The virtual cyber world can also expand the operational space. The virtual world exists in parallel with the physical, bifurcating the cognitive domain. Networks can connect disparate elements of a group or a cause, expanding their presence over borders or even continents, making their containment practically impossible. The cyber domain will also intersect with advancements in sensor developments, AI and machine learning. Vast quantities of data can be increasingly amassed with ease, given more capable and more cost-effective sensors. AI will allow these data to be processed and interpreted, while machine learning—a subset of AI—will enable systems to identify patterns, make decisions and enhance themselves through experience and information.

The volume of data is already becoming unmanageable and will increasingly rely on such techniques to make it intelligible. In 2011, the collective fleet of US unmanned aerial vehicles (UAVs) amassed 327,000 hours, or 37 years' worth of footage—more than could be screened, much less characterised.[4] AI and machine learning will be the only way to effectively sift through these mountains of data. But increasingly, as capabilities evolve, these systems could be tasked further to operate without human supervision, ultimately supporting or even directing the command and control of war.

As AI systems may be able to make decisions much faster, depending on the context parameters, command and control may necessarily be handed over, as human reactions may not be sufficient. AI has already been proven successful in air-to-air combat tests run by the US Defense Advanced Research Projects Agency (DARPA). While AI systems are beginning to create tactics and manoeuvres superior to those of human actors, experiments are also indicating that they will be able to support decision-making further up at the operational level. The ability to process large volumes of data on common military variables such as enemy positions, weapon capability and

---

4   'Artificial Intelligence Is Changing Every Aspect of War', *Battle Algorithm* (blog), *The Economist*, 7 September 2019, www.economist.com/science-and-technology/2019/09/07/artificial-intelligence-is-changing-every-aspect-of-war.

ranges, terrain and climate will enable the system to infer predictions in ways much faster than traditional estimates. Ultimately, such systems may develop a strategic rationale.

As indicated by the chapters in this volume on remote and autonomous platforms, there is an urgent need to ensure that such systems operate in accordance with human values and the laws of armed conflict. In the near future, the act of killing will remain a human function, but it may not always. Factors such as proportionality and target discretion will need to be incorporated. These capabilities will need to be able to cope with the pace and fog of the battlefield, and potential efforts to hack into or spoof the system. The consequences of failure in any one of these parameters will be dire. However, developments are progressing more quickly than societies' ability to incorporate these systems into doctrine or an ethical framework, threatening our ability to maintain control of this evolution.

Developing technology and outer space

AI technologies will aid the realisation of the potential of the space domain. The miniaturisation of sensors and their reducing cost, as well as the reduction of cost to access space, generally point to an ever-increasing capability to generate data on virtually every aspect of human activity. Satellites can be manufactured at a fraction of the cost that would have been required at the end of the Cold War, and of a size that enables large numbers to be launched per mission. This could result in space-derived data sets providing meta-calculations of economic or military activity on a scale unimaginable until recently. But this would remain unrealised without processes of automation.

In tandem with this—in fact, driving these processes—is a separate trend that is reflected in space, possibly more than any other domain: the rise of the commercial sector. Once the preserve of military operations, and shrouded in a veil of classification, the space domain's complexion has irrevocably changed. Much of the most advanced research and development is now undertaken in commercial labs. So dominant is this trend that more than three-quarters of the space economy is commercial, and the largest spacefaring entity by number of satellites is a commercial operator— SpaceX. If information is power, then that baton is being handed over to the commercial sector.

This reflects a key theme in 4IR: the transparency of information. No longer can the information derived from space be restricted, nor should it be, and this will enable exponential growth in information, generating a myriad of new applications and avenues of exploration. It will, however, also change military uses of space, where expertise is held beyond those institutions, and operational secrecy is no longer possible.

However, the critical nature of space ensures that it does not escape the tensions of major states, and the spacefaring powers will continue to seek advantage in and from the domain. On the eve of the conference which generated this volume, India tested an anti-satellite missile (ASAT), and Russia carried out a kinetic intercept of their Nudol missile in late 2021, signalling a rise in competitive behaviour. Any spillover into outright conflict in space could threaten our ability to access the domain, as even a single attack creates debris and effects that have the potential to impact all users, and thus the collective ability to exploit the burgeoning services and information. Such concerns have until recently mostly escaped the policy-making milieu. The dependence on space and the vulnerability of those systems is only now becoming apparent to many.

But while space quite clearly expands the operational scope of modern war, questions remain as to what extent this will continue. What should the purview of a military space force be? Should this go beyond the orbital system of Earth, to encompass lunar orbit? To what extent should a nation's space force follow commercial trade into deep space? The possibility exists then of a literally infinite expansion of the operational space.

It might also be argued that the recent ASAT tests noted above have shifted the perceptions about humanity's connection with space. While distant, and in some respects seemingly virtual, the recent Russian ASAT test created the need for the crew of the International Space Station, which included Russian cosmonauts, to shelter. This might be seen as an important step in the realisation that action in space, or into space, can impact human lives. The connection between Earth and space was made far more apparent, perhaps bolstered by the parallel space tourism flights. With this, our sense of place may change.

Shifts in society, resources and the distribution of power

While there is a great deal of emphasis on technological development in predictions of future conflict, the advances in capabilities are only one aspect. The distribution of technology is also likely set to change, and this will affect the advantages that Western states have historically held. The rise of commercial research and development has already been noted above, but in addition, adversaries, from non-state actors to other states, have access to more of the same technologies that had been the preserve of advanced economies. This increasingly levels the playing field and allows those very technologies to be used against Western states. Today, UAVs are being produced and widely exported by states well beyond the traditional manufacturers of military aerospace systems. For example, Turkish UAV designs were used by Azerbaijan in the Nagarno–Karabakh conflict, introducing remotely piloted aircraft into a conventional anti-armour role for the first time. Cyber capabilities are no longer the exclusive preserve of the Western states that developed them, and their offensive use can be conducted by other states, terrorist groups or even individuals. Various social media outlets have become conduits for sophisticated interference with political processes, by manipulating information and thus opinion, creating stark division within populations, effectively allowing foreign powers or groups a form of direct access. The facets of free speech and openness inherent in democracies make them especially vulnerable to cyber and internet threats, providing the prospect of significant and widescale impact while presenting minimal risk to the originator.

Wider changes in population growth, demographics and pressures on resources will further complicate the security context. Overall population growth is set to rise approximately 20 per cent to 9.7 billion by 2050. While significant in itself, this growth will not be uniform, with demographic factors such as life expectancy, the size of a population's economically productive sector and migration creating variations globally. A broad and consistent trend within this is urbanisation, with predictions that by 2050 the percentage of the global population in urban centres will have risen from a little over half in 2015 to over two-thirds in 2050.[5] Highly concentrated in small areas, driving the majority of energy use, urban centres also contain

---

5   United Nations, *World Urbanization Prospects: The 2014 Revision* (New York: United Nations, 2014), doi.org/10.18356/527e5125-en.

great social inequality. The combination of these factors creates a number of pressures and tensions that can undermine the stability of urban centres, and where military response may be necessary, these environments present a challenging landscape.

The growth of population and the impact of climate change are also creating pressures on resources. These include basic staples to support populations, such as water, certain food commodities and fuel, as well as strategic materials that are fundamental for industry and technology. Water security is often viewed as the most critical and fundamental, increasingly affecting developing areas most severely. As pressures on this resource increase, it presents another axis of instability. Lack of access may trigger migration, while interference in water resources is an instrument of coercion available to inimical actors.

Fragile supply lines of oil and gas also present vulnerabilities. Suppliers can threaten to cease supply or reroute supply lines in a way that disadvantages other target states. Europe is vulnerable to Russian decisions in this regard, for instance. Deliberate attacks can have an immediate effect on supply and prices, as was the case with the UAV strikes against Saudi production facilities. Tenuous fuel supply routes can also be a vulnerable point of a military operation, increasing cost and complexity. This was a constant problem in the North Atlantic Treaty Organization (NATO) campaign in Afghanistan, where the cost of getting fuel to troops could be up to $400 per gallon. While the development of new sources of energy may change this vulnerability, it will of course create new shifts as certain states seek to dominate the market for new energy materials and technologies.

New materials required for the manufacture of high-tech products, ranging from military systems to mobile phones and electric cars, have also become a choke point in related supply lines. Rare-earth elements (REEs) or rare-earth metals (REMs) have become a critical ingredient in global industries related to many new technologies, and an extremely high proportion of these are controlled by China. This near-monopoly, combined with escalating demand, provides suppliers with immense geopolitical leverage over consuming states. While it remains to be seen how market forces might shape these advantages, and whether alternative sources or indeed materials may be created, there is sufficient potential for this to remain a facet of leverage, and possibly an avenue through which to challenge the distribution of power.

***

This volume has examined a wide spectrum of factors that will shape the way conflict is waged in the future. The impact of technology and new domains, and the shifting landscape of society will combine to increase the level of complexity in future war, driving ever-increasing uncertainty. While technological change has been a constant over centuries, the current rate of progress is viewed as closely interconnected and evolving exponentially. Fields such as robotics, advanced materials development, genetic modifications, the Internet of Things, drones, autonomous vehicles, neurotechnologies, sensor technologies, cyber capabilities and the related fields of AI and machine learning are becoming more integrated, fusing the physical, digital and biological spheres. As articulated by the concept of a fourth industrial revolution, this confluence provides boundless potential for change. These advances will continue to interact with developments resulting from regional population growth, the effects of climate change and the competition for finite resources. To oversee this remarkable advance, ethical frameworks will need to be developed, though the scale and speed of change will test society's ability to keep pace.

These pressures will continue to shape society, and how war is waged. The shape of the future battlespace will be extended into new physical and virtual domains, while the time in which to respond to threats and actions will be simultaneously compressed. In addition, fundamental components of the current security environment, such as the primacy of the state, may be increasingly challenged.

While these changes may be considered to impact the character of war, it might also be questioned as to whether the fundamental nature of war will be altered. If AI and machine learning increasingly challenge humans in the cognitive domain, and wars are increasingly waged by non-biological intelligence, will war remain a human activity?

References

'Artificial Intelligence Is Changing Every Aspect of War'. *Battle Algorithm* (blog), *The Economist*, 7 September 2019, www.economist.com/science-and-technology/2019/09/07/artificial-intelligence-is-changing-every-aspect-of-war.

Deptula, Brigadier General David A. *Effects-Based Operations: Change in the Nature of Warfare*. Arlington: Aerospace Education Foundation, 2001.

Falls, Cyril. *A Hundred Years of War, 1850–1950.* London: Duckworth, 1953.

Schwab, Klaus. 'The Fourth Industrial Revolution: What It Means, How to Respond'. *World Economic Forum*, 14 January 2016, www.weforum.org/agenda/2016/01/the-fourth-industrial-revolution-what-it-means-and-how-to-respond/.

United Nations. *World Urbanization Prospects: The 2014 Revision.* New York: United Nations, 2014. doi.org/10.18356/527e5125-en.

Index

Note: page numbers in italics indicate information found in tables, diagrams or other illustrative material.